# Rethinking Bioassessment of Freshwater Ecosystem Health

Robert C. Bailey

# Rethinking Bioassessment of Freshwater Ecosystem Health

Springer

Robert C. Bailey
Ontario Tech University
Oshawa, Canada

ISBN 978-3-032-12426-5 ISBN 978-3-032-12427-2 (eBook)
https://doi.org/10.1007/978-3-032-12427-2



This Springer imprint is published by the registered company Springer Nature Switzerland AG
The registered company address is: Gewerbestrasse 11, 6330 Cham, Switzerland

# Preface

Any author of a scientific monograph with a bit of humility (I have a little bit) begins with a justification for the work…why yet another treatise on [*insert subject here*]? I started thinking about doing this project in 2014 when I was finishing up five years as a university administrator with lots of ideas percolating about how to improve bioassessments of freshwater ecosystems. Ten years earlier, Trefor Reynoldson, the late Richard Norris, and I had published a book on the Reference Condition Approach (RCA) to bioassessment (Bailey et al. 2004), and Simon Linke, Adam Yates, and I had just edited a special issue of Freshwater Science where several of our colleagues compared analytical approaches to doing RCA assessments using three datasets (Bailey et al. 2014). Life and work got in the way of that plan, as I accepted another appointment in administration that occupied most of my time until 2019. This was a blessing in disguise, because as I began the final chapter of a long academic career, I read a few things and talked to some people and found my whole approach to bioassessment changing. This book became the experience of my "rethinking" of bioassessment.

So what, exactly, did I "rethink"? First of all, I thought that those of us working on bioassessment, including both researchers and practitioners, should be much more careful about how we describe both the natural and human activity environment of the ecosystems we assess. I felt that this was caused by the lack of a carefully considered conceptual model. Being more thoughtful about developing a clear conceptual model that showed my view of how the natural and human activity environments affected the biota in ecosystems had helped me to be more precise about defining Reference versus Test ecosystems, another aspect of bioassessment that is surprisingly vague. If we clearly define Reference sites, we can use them to model the relationship between the natural environment and the biota at those sites, and then apply the resulting model to assess Test sites. We can then use *all* the sites (Test and Reference) to diagnose whether or not and how specific kinds of human activity are affecting the biota in ecosystems.

To build statistical models, I also soundly rejected (no pun intended) what is known as frequentist, null hypothesis significance testing (NHST) in favor of a Bayesian modeling approach. The legendary Richard McElreath's book (McElreath 2020) and incredible online courses got me on this path. For me this was not a philosophical preference; I really like the infinite array of alternatives that Bayesian fitting checks out, along with their likelihoods, much more than NHST's null versus alternative hypothesis dichotomy. I also like the way that Bayesian fitting can be continuously refined with newly collected data. After decades of using and teaching "stats", I have embraced this new way of analyzing data and I haven't regretted it. I should emphasize, though, that most of the ideas in this book … the importance of the conceptual model, defining Reference versus Test sites, and building Reference Condition and diagnosis models … don't depend on using a Bayesian approach.

There are a lot of very useful books and papers on bioassessment of freshwater ecosystems, some of which I cite in this book. This is not a review of all the great and not-so-great bioassessment methods developed and applied over the years. I try to show clearly how my approach works with both a simulated and real dataset, and I provide R scripts at the end of each chapter where they're used so you can try it too. But I don't compare this approach to what others do (or what I used to do) because it's too much of an apples-to-oranges situation. I don't expect to change the bioassessment world with my rethinking, but hopefully this forms the basis of some constructive discussions. At the very least, I hope that other folks are a little more careful in building a conceptual model for their program, the definition of Reference versus Test sites, the selection of candidate predictors they use in Reference Condition and diagnosis models, and how they actually do and interpret the bioassessment.

A lot of people have contributed to my reflections and work on bioassessment science over the years. Like many academics, my long-ago but never-ending membership in my Ph.D. supervisor's group was the initial spark. The (Roger) Green lab provided an inspiration that has never waned, and I am lucky enough to live close enough to Roger that we still occasionally get together to argue about bioassessment, among other issues. The late, great Richard Norris regularly rears his virtual head when I'm thinking about bioassessment, and I still collaborate from time to time with Richard's and my long-time co-author Trefor Reynoldson. John Bailey, my brother and lifelong friend (at least since we stopped fighting after high school) helped me bring theoretical ideas about bioassessment to reality and application in the work we did on Yukon River Basin streams. Adam Yates and Simon Linke, former grad students in my lab who have gone on to contribute so much to their own research areas, are always happy to listen to my ravings and put me back on course when I wander. Colleague and old friend (and another Green lab alumnus) Lee Grapentine at Environment and Climate Change Canada generously shared the GLBEAST dataset along with his insights.

I am deeply indebted to Jan Ciborowski for reading a late draft of this little book, and gently but firmly challenging my thinking and clarity of communication. After I worked through Jan's very useful comments, aquatic scientists Bob Brua and Adam Yates reviewed the penultimate draft. I didn't always take Jan's, Bob's, and Adam's advice, although I probably should have, and therefore I remain responsible for any remaining clunkers in the text.

Lastly, I dedicate this work to my partner through all of the twists and turns of my life in science and academia. We got married a couple of months before I started my MSc and she keeps me just as grounded now, more than 45 years later, as she did then. Thanks Chris!

Kingston, ON, Canada
June 2025

Robert C. Bailey

## References

Bailey, R.C., Linke, S. and Yates, A.G.: Bioassessment of freshwater ecosystems using the Reference Condition Approach: comparing established and new methods with common data sets. Freshwater Science **33**, 1204–1211 (2014)

Bailey, R.C., Norris, R.H. and Reynoldson, T.B.: Bioassessment of Freshwater Ecosystems-Using the Reference Condition Approach. Kluwer Academic Publishers, Boston (2004)

McElreath, R.: Statistical Rethinking: A Bayesian course with examples in R and STAN. 2nd edition. Chapman & Hall/CRC (2020)

# Contents

# Chapter 1
# Start the Journey by Looking Back

I got started as an academic in the 1980s with my Bachelor's degree in Ecology in 1980, an MSc in 1982 focused on fingernail clams, and my PhD in 1987 studying freshwater mussels. By the end of the 1980s, I was a junior professor at Western University in London, Ontario, one of Canada's long-established "ivy-leaguish" universities, with prominent law, medical, and engineering schools. I was, and remain, a freshwater ecologist…never really considered quite as "hard science" as my genetics or developmental biology colleagues or as theoretically grounded as my evolutionary biology friends.

In my early research years, I was mainly interested in freshwater molluscs, especially whether or not the variation in their shells was due to "nature or nurture" (Bailey and Green 1988). But later work in my lab that really resonated with colleagues and people beyond the university was research on bioassessment. Everywhere from international scientific conferences to family holiday dinners, I talked about bioassessment…assessing the effects of human activity on an ecosystem by studying the organisms that are living there. Since in freshwater ecology we often study the invertebrates in the ecosystem visible to the naked eye (*aka* macroinvertebrates), I would regularly show field course students or my kids' grade school classes the sensitive and tolerant "bugs" in a local stream or lake.

Most lay people are quick to get the idea of bioassessment but have trouble seeing it as complex science. If the "canary in the coal mine" approach works so well, are there really any difficult research questions left? It's true that you don't need intensive sampling and sophisticated modeling and statistical analysis to distinguish a wilderness stream from a heavily polluted drainage ditch adjacent to an industrial-scale chicken farm. But the sort of questions we ask in bioassessment are often more subtle. Is it better to have a 50-m or a 150-m buffer zone of "no logging" along a creek to protect the stream ecosystem? Are the "best management practices" used by a farmer to protect streams draining the farm doing any good? Have lakes in a remote, boreal forest area been affected by climate change? Is the third condo on the shore of this lake going to be a tipping point in the ecosystem health of the waterfront?



R. C. Bailey, *Rethinking Bioassessment of Freshwater Ecosystem Health*,
https://doi.org/10.1007/978-3-032-12427-2_1

As mentioned, the logic of bioassessment is deceptively simple: What do the organisms living in an ecosystem tell us about its health? Underlying this simplicity are two important presumptions:

1. The biota temporally and spatially integrate often transient changes in the physical and chemical environment, and therefore are a more **consistent and reliable signal** of significant ecosystem change than simply measuring the chemical or physical properties of the water.
2. Changes in the physical and chemical environment that don't affect the biota are **unimportant**.

If the only species present are those tolerant to a pollutant or stressor present in the environment, it would be fair to conclude that the pollutant has degraded the health of the ecosystem. Conversely, if species that are sensitive to the pollutant are present in a similar, nearby ecosystem where the pollutant is absent or has been reduced or eliminated, we can reasonably infer that the ecosystem is healthier because of the absence or reduction of the pollutant. It is not just the presence of the pollutant that is used to judge ecosystem health…it is its specific effects on the biota in the ecosystem.

As I discuss below, there have been many significant advances in the bioassessment of freshwater ecosystems over the past 50 years. With all this progress, though, an actual working definition of a healthy ecosystem (and, therefore, an *unhealthy* ecosystem) remains somewhat murky and muddled. Many ecologists developing and applying techniques to assess the effects of human activity on ecosystems consider Aldo Leopold's Sand County Almanac (Leopold and Schwartz 1949) to be the first clear articulation of a land "ethic". Leopold famously stated that land use "…is right when it tends to preserve the integrity, stability, and beauty of the biotic community. It is wrong when it tends otherwise." But this inspiring quote is fraught with value-laden ("beauty") and non-operationally defined ("integrity", "stability") terms that continue to vex environmental assessment to this day. Rohwer and Marris (2021) argued that "ecosystem integrity [and by extension, ecosystem health] is neither real nor valuable". If they're right, the consequences for bioassessment programs are profound…programs based on a fuzzy definition of a healthy ecosystem are doomed to be just a way of describing status and trends of the ecosystems studied and lead to endless battles about the effects of existing or proposed human activity. Karr et al. (2021) responded passionately and effectively to Rohwer and Marris (2021), but they fell short in describing operational details while tackling the bigger philosophical issues on the table.

Yates et al. (2019a) suggest a different approach, where ecosystem services are considered as part of defining a healthy ecosystem that is in "Reference Condition", and Stoddard et al. (2006) have a great discussion of the Reference Condition…the biota we expect to see in sites that have not been degraded by human activity. But with all of this great work and discussion, I'm still frustrated by a pervasive lack of clarity in how we define Reference sites and at least *relatively* healthy ecosystems when setting up a new bioassessment program. This has become crucially important as freshwater ecosystems around the world have seen precipitous declines in biodiversity because of human activity (Maasri et al. 2022).

The biota in lakes and streams respond to and influence the dynamics of the chemical and physical environment. For many millennia, people living on the land have carefully observed freshwater ecosystems, keenly aware of the material and spiritual goods and services they provide, including water for drinking and transporting waste, and a habitat for other organisms they rely on to live. More recently, Western science has developed tools to assess the state of ecosystems and assemble circumstantial evidence for human causes of this state. Bioassessment of freshwater ecosystems in Western science has a long, well-reviewed history (e.g., Hynes 1966; Metcalfe and Reynoldson 1989; Bonada et al. 2006; Buss et al. 2015), and many of the ideas I develop here are based on work cited in these reviews and other thoughtful work (e.g., Norton et al. 2015). Rather than summarizing these, I will try to capture the essence of the evolution of bioassessment as a series of overlapping and interacting eras, particularly as it relates to the approach I present in this book. Of course, no timeline is as neat and tidy as it may look in retrospect…many biomonitoring and bioassessment programs are carried out today much as those done 100 years ago!

## 1800s—1950: Early Years—Applied Limnology

In the early twentieth century, German aquatic scientists Richard Kolkwitz and Maximilian Marssdon published the Saprobiensystem (English translation in (Kolkwitz and Marrson 1967b, a), which is often cited as the foundation of bioassessment, at least from a formalized, Western science perspective. They list species including protists, invertebrates, and fishes that defined sequential zones downstream of a sewage outfall. Their logic is consistent with the basic premise of bioassessment I've described; the biota that are present reflect their environment, including properties like the amount of dissolved oxygen, which is affected by the decomposition of waste. As I reread Kolkwitz and Marrsdon today, I am struck by how they elegantly link the ability of a particular group of "animal and plant" species (some organisms were from other Kingdoms) to survive and thrive in an environment and then alter that environment to the benefit of some and the detriment of others.

Understanding the biology of the biota in ecosystems variably exposed to pollutants (e.g., which species thrive and which don't) is key to building the circumstantial case underlying bioassessment. If the community composition of biota did not change among zones varying distances downstream of a sewage outfall, or they changed for reasons other than the outfall (e.g. differences in flow rate or substrate), then the case is weakened. That much is still true, all these years later, and sometimes forgotten with more sophisticated modeling techniques that tend to be further and further from the data. Even before this seminal work, early limnologists in many ways defined what became community and ecosystem ecology, and described relatively well-demarcated lake ecosystems that were classified as either healthy or degraded by various types of human activity (Egerton 2014). A deep, albeit descriptive knowledge of freshwater ecosystems in their more or less natural state led to a greater understanding of how the biota were different in unhealthy ecosystems. This is perhaps best characterized by Canadian limnologist DS Rawson, who Egerton (2016) notes was the first to classify lakes as eutrophic, mesotrophic, or oligotrophic (Rawson 1939), and was one of the first aquatic scientists to explicitly propose a conceptual model of how both natural features and human activity determine the trophic status of a freshwater ecosystem.

## 1950s—1980s: Midlife Crisis—from Indices to BACI

In the middle twentieth century, ecology (including applied ecology) became more quantitatively and theoretically grounded. Ruth Patrick's work on diatom assemblages in rivers exposed to varying levels of stressors (Patrick 1963) was clearly based on the logic of bioassessment but also had a strong link to the Island Biogeography theory initially developed by MacArthur and Wilson (1963). Information theory (Shannon 1948) provided a "diversity index" that combined the number of taxa in an ecosystem with the "evenness" of their abundance or biomass. By Shannon's measure, not only was a community with more species more diverse, but those with a given number of species and a more evenly distributed abundance among the species were "more diverse" than communities dominated by just a few of the taxa present. Other diversity indices were developed, with most having no requirement of knowing anything about the biology of the taxa or even their identity…just that they were distinct from one another, at least morphologically. Several of these, and the simplest and most robust measure…the number of taxa (**taxon richness**)… are a part of many bioassessments now, and **evenness** (the balance among taxa in abundance) is often reported as well. The often unstated assumption was and is that higher diversity is "good"… and any human activity that reduces it is "bad". Of course, in many situations (including both examples we explore later in the book), humans can increase diversity of the biota in ecosystems that tend to have lower diversity in their natural state (e.g., some oligotrophic lakes where humans have added nutrients).

Along with the popular use of diversity indices in bioassessments, other indices incorporating relevant biological information about the taxa were also developed. These were usually calculated as an abundance-weighted average of each taxon's sensitivity or tolerance to one or more pollutants, which was often determined from observational data. Hilsenhoff's Biotic Index (Hilsenhoff 1987) in the US, and the Biological Monitoring Working Party (BMWP) initiative in the UK (Walley and Hawkes 1996) are still used frequently, and combining such indices (along with diversity indices and other metrics) into so- called multi-metric indices (Karr 1987) has become common practice, especially in many US bioassessment programs (e.g., Barbour et al. 1999). As with simple measures of diversity, these indices all have an implicit "polarity", assessing ecosystem health on an absolute rather than relative scale…higher diversity and a lower tolerance (or higher sensitivity) score are indicative of greater ecosystem health.

Indices came to dominate bioassessments as descriptors of the biota, but it wasn't long before statistically minded ecologists also started carefully considering what they considered the "proper" design and analysis of such studies. This was epitomized by Green (1979) with his development of the famous Before-After-Control-Impact (BACI) design, still popular as a kind of gold standard of environmental assessment 50 years later. Some version of the BACI design has been required in most environmental impact assessments since the 1970s (e.g., Somers et al. 2018) and much has been done to both critique (e.g., Hurlbert 1984) and extend (e.g., Underwood 1992) the BACI design, as well as the associated Null Hypothesis Significance Test (NHST) paradigm.

The logic of Green's BACI design has always appealed to the statistically astute environmental scientist assessing human activity, whether negative "degradation" or, more rarely, positive "rehabilitation". In Green's paradigm, the "Control" site is upstream of (or away from) a hypothesized stressor, while the "Impact" site is just downstream of (or adjacent to) the stressor. In the simplest Control versus Impact (CI) situation, where the stressor already exists (e.g. the sewage treatment plant is operational), the bioassessment compares the Control and Impact sites, and any difference between them is considered evidence of an effect of the stressor (Fig. 1.1).

Although the circumstantial evidence of a CI design is quite weak…there could be many causes of differences in the biota between the Control and Impact sites…even now this simple comparison forms the basis of many bioassessments. It's not great, but it's better than nothing, and such evidence can be bolstered (or refuted) by looking carefully at *how* the biota differ between the Control and Impact sites. Are they different in a way that is consistent with the hypothesized stressor's known effects on the ecosystem?

The real power of the true BACI design comes into play when change over time at both the Control and Impact sites is incorporated into the study design (Fig. 1.2). Sometimes a project is planned but not yet executed (e.g., a proposed new sewage treatment plant), so sampling of both the Control and Impact ecosystems can happen prior to the presence of the new stressors. In this case there are four sets of samples

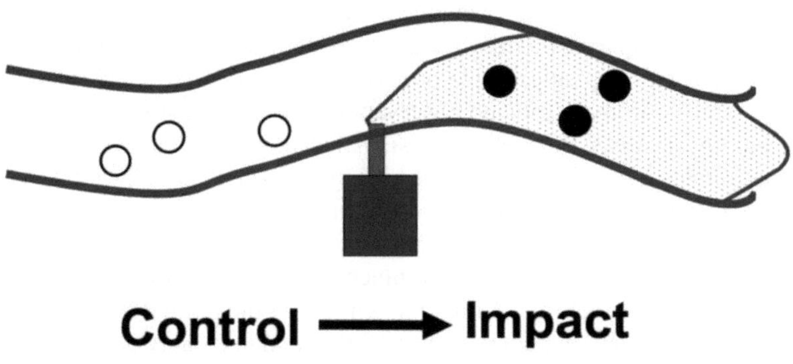


**Fig. 1.1** CI bioassessment design comparing biota in Control (open circles) *versus* Impact (closed circles) sites

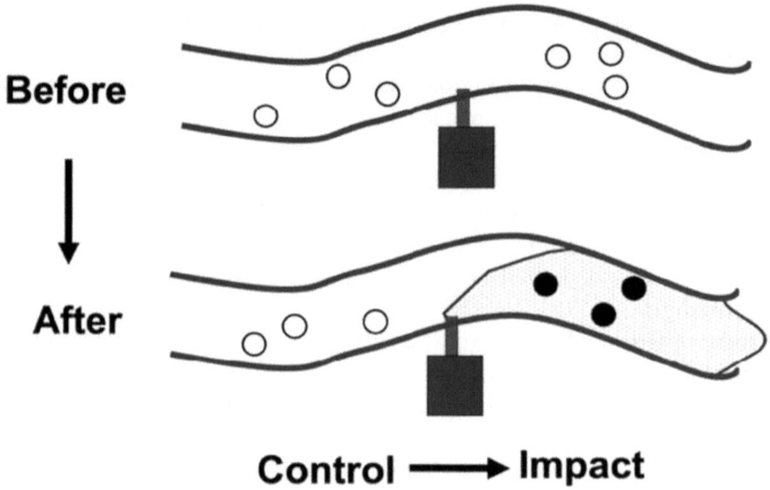


**Fig. 1.2** BACI bioassessment design comparing biota in Control (C) versus Impact (I) sites Before (B) and After (A) the human activity has been initiated

taken…both Control and Impact sites are sampled "Before" the stressor is turned on, and both areas are sampled again "After" the stressor is operational. *Any change* observed at the Impact site between Before and After that is *not observed* at the Control site over the same time period is considered strong (or at least *stronger* than CI) evidence of an effect of the stressor on the ecosystem.

Although evidence from a BACI bioassessment is much more compelling than the simple CI comparison, it's still important to look at the difference in Before versus After changes and make sure they are consistent with the known effects of the stressor on the biota. And of course there is just *one* Control and *one* Impact site, with many possible reasons having nothing to do with the stressor that the dynamics of one area are different than those of the other. This critical, *scientific* constraint of the BACI design tends to get lost behind a *statistical* constraint true of all observational studies…taking multiple benthic samples in both the Control and Impact areas and then statistically comparing them, whether or not Before and After samples are taken, came to be known as a great example of *pseudoreplication* (Hurlbert 1984).

It is certainly true that replicate observations are not randomly assigned to the Control and Impact areas. But beyond the statistical problem is the overly constrained conceptual model. The attractiveness of having two sites that only differ (or to be fair, *mainly* differ) in their exposure to some human activity captivates assessment scientists, but this comparison doesn't consider how several Control sites unexposed to the human activity might vary, and how much the degree of difference in the Impact site compares to that variability. I think recognition of the limited inference possible with just one Control and Impact site captures the fundamental BACI problem better than framing it as a statistical challenge (e.g., Stewart-Oaten et al. 1986). This is an often unrecognized reason for the subsequent success of the Reference Condition Approach (RCA) to bioassessment described below.

The significance to bioassessment of the BACI design also extends far beyond the inferential strength and challenges of the design. Perhaps most importantly, in judging the effects of stressors on ecosystem health by comparing Control and Impact sites, BACI moved assessment from an **absolute** to a **relative** scale. Instead of judging whether diversity was high enough or a tolerance score was low enough for a healthy ecosystem, decisions were based only on the comparison…is diversity *different* at the Impact site than it is at the Control site? This was stronger statistical evidence, but reduced the responsibility of the scientist for asserting what the necessary diversity or tolerance score was for a healthy ecosystem. In some ways the NHST stats behind the BACI design allowed abdication of the responsibility of the bioassessment practitioner as to what was an ecologically important difference in the biota, and led to lots of angst and arguments about the power of the hypothesis tests to this day. Related to this challenge, there was increasing awareness that we need to be designing bioassessments to detect **cumulative effects** of a variety of stressors in more realistic and complex settings than the classic BACI design (Underwood 1992). If the ecosystem is already at least partially degraded upstream of the Control site, there might be no discernible difference between it and the Impact site.

At the same time as the growing popularity of the BACI design, there was an increasingly critical view of "indices", particularly those (like diversity indices) that were not linked to the biology of the community members. Many thought indices missed the subtleties of changes in the structure and composition of the biota in the face of human-caused stressors (e.g., Green 1979). The "community matrix" (abundance or biomass of each taxon in each ecosystem sampled) quickly became rather uncritically accepted as the best way to describe the biota in an ecosystem and evaluate ecosystem health. There are two important challenges to this that have a profound effect on bioassessment programs to this day. First, "the sampling problem", as carefully considered at the time by several authors, most notably Elliott (1977), Resh (1979), and Downing (1979), challenges blithe acceptance of the community matrix as a precise quantitative description of the biota in an ecosystem. Recently, many bioassessment scientists (e.g. Baird et al. 2016) have concluded that only presence/absence data (or, more accurately, **taxa detected/undetected** data) are estimated well enough for bioassessment analysis. This is the approach I use later in the book in looking at the GLBEAST dataset from sampling nearshore ecosystems on the Canadian side of the North American Great Lakes.

The second challenge of the community matrix is uncritical acceptance of the morphology-based taxonomy of the day being the *best* descriptor of the biota. The last 20 years have seen the development and application in biomonitoring and bioassessment programs of "environmental DNA" (eDNA) to better characterize the biota present in an ecosystem (e.g. Baird and Hajibabaei 2012, Hajibabaei et al. 2016, Maasri et al. 2022). Other descriptors from more recent work on metabolomics and fatty acid analysis (e.g., Whorley and Wehr 2018) looks very promising. There is also evidence that a traits-based approach (e.g., Orlofske and Baird 2014) to describing biota may provide a more complete picture of what array of organisms are living in the ecosystem, and thus been a useful addition to many bioassessments. It too relies on more accurate taxonomic descriptions of the community than morphology-based identification can provide.

## 1980s—2020s: A Matter of Scale—the Reference Condition Approach

As discussed, the BACI bioassessment design developed by Green (1979) was based on the notion of *relative* ecosystem health: How did the biota in what Green called a Control area differ from those in an area exposed to human activity (the so-called Impact area)? Most criticisms of the design were focused on its statistical aspects rather than the narrow definition of the Control ecosystem. The **Reference Condition Approach** (RCA) essentially extended BACI by characterizing the biota in a geographically and ecologically diverse set of Control sites that were relatively unexposed to human activity. This defined the Control or **Reference Condition** against which the biota in ecosystems in "Impact" (or more properly **Test**) areas were compared (Fig. 1.3). Statistically, rather than the simple comparative model of the BACI design, RCA models exploit the relationship between the biota and its natural environment in Reference ecosystems to "predict" what biota should be present in a Test ecosystem if it's in **Reference Condition** (i.e. has good ecosystem health) and then statistically evaluate the difference between the prediction and what has been observed.

Predictive models based on the Reference Condition Approach were first developed for, and applied to, national biological monitoring programs in the United Kingdom (UK) beginning with a project in 1977 at the River Laboratory of the Institute for Freshwater Ecology (IFE). The project had two objectives (Wright et al. 1984): classify "unpolluted" running-water sites based on benthic macroinvertebrates, and determine whether the biota expected at unstressed sites could be predicted from physical and chemical features of the site. This later evolved into the massive River Invertebrate Prediction and Classification System (RIVPACS) program that continues to this day and has spawned programs around the world, including in Canada (Strachan and Reynoldson 2014), Australia (Nichols et al. 2014), and various countries in the EU (Costas et al. 2018).

Although analysis methods vary substantially, all RCA assessments are based on measuring the relationship between the biota and its natural environment at Reference sites (those less exposed to human activity) and using this relationship to construct a model that predicts the expected biota at a Test site (those more exposed to human activity). The deviation between the expected biota and what is actually found at the Test site is used as a measure of how much and how ecosystem health has been affected by human activity.

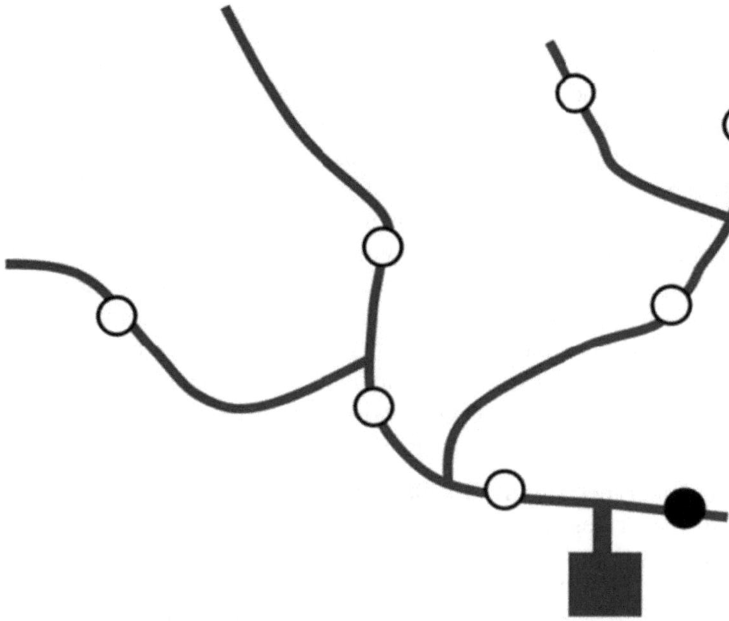

**Fig. 1.3** RCA bioassessment design comparing biota at a Test site (closed circle downstream of human activity) to a set of Reference sites (open circles unexposed to human activity)

## Why Rethink Bioassessment?

I've researched different aspects of bioassessment and the Reference Condition Approach for more than 30 years, and helped implement the RCA in everything from regional to national scale programs. In the last several years, I have looked very critically at every aspect of the approach in an effort to increase its efficiency, effectiveness, and utility as a management tool. In laying out my "rethinking" in this book, I don't spend any time comparing the performance of my approach to others, because to me there is no benefit in criticizing past work (including my own) when it is difficult or impossible to compare performance on some kind of standard scale. My rethinking has focused on several areas of bioassessment where I think real progress can be made regardless of the particular analytical method used to carry out the bioassessment:

1. **Building the Conceptual Model**: One thing I have come to recognize is the overarching importance of building a conceptual model of the system you're assessing. I've also looked or asked for such a conceptual model when I hear or read about bioassessment programs, and usually get a combination of two responses: "Of course we have an implicit conceptual model" or "We'd like the data to tell us what's going on". The problem with this sort of response is that

the environmental descriptors used in defining reference sites or modeling the reference condition may be problematic for one or more of the following reasons:

- descriptors of the natural environment are affected by human activity (e.g. sediment particle size)
- descriptors of the natural environment are not capturing the aspect of the environment intended (e.g. latitude and longitude)
- descriptors of the human activity environment are a result rather than a measure of human activity (e.g. total nitrogen).

  A very useful environmental descriptor might be missed because of lack of access to either the information or the technology to get the information. A good conceptual model will identify gaps like this that need attention. The conceptual model doesn't include numbers or even signs of the effects of these hypothesized causes, but just lays existing knowledge on the table and can be a great guide as to what and what not to measure to describe the environment.

2. **Defining Reference versus Test sites**: Adam Yates and I have long been advocating for better description of both the natural (Yates and Bailey 2010b) and human activity (Yates and Bailey 2010a) environments at each site for their proper classification into Reference *versus* Test categories. Defining each of these sets of environmental descriptors (based on the conceptual model) can be helpful in both designing a bioassessment program (e.g. making sure that the full range of natural environments in the program area get characterized) and the definition of the human activity "boundary" between Reference and Test sites. This part of an assessment program has nothing to do with the biota…it is all about the range of natural and human activity environments that are present and used in our modeling.
3. **Modeling the effect of the natural environment on the biota in Reference sites.** I think development of these so-called Reference Condition models has been stymied by lack of a clear conceptual model and too many candidate predictors, including some affected by human activity and some that are confounded with others. There has been much more work done on the modelling technique (e.g., biota-based classification followed by discriminant analysis using natural environment descriptors) than either the conceptual model or the careful selection of which candidate predictors are considered.
4. **Assessing Test sites using Reference Condition models.** If the natural environment of a Test site is "plugged in" to a Reference Condition model, comparing the biota predicted from the model to what is actually found at the Test site is the crux of the bioassessment. There has been a lot of work done on appropriate statistical tests of "O:E" scores and other measures of the difference between what is expected and what is observed at a Test site. I think a Bayesian fitting approach may be more useful and appropriate than the NHST approach usually used for the bioassessment.
5. **Modeling the effect of human activity on Reference and Test sites**: After a bioassessment is done, there is always interest in why a Test site doesn't have the biota expected if it is in Reference Condition. Usually there is an attempt

to correlate a variety of stressors with bioassessment results to build a circumstantial case. With the example datasets, I'll demonstrate what I call **diagnosis** models, where I model variation in the biota as caused by both the natural and human activity environment and look for evidence of human activity effects either independent of or interacting with the natural environment of the site.

6. **Identifying and interpreting temporal trends in Reference and Test sites**: "Time" (other than "Before versus After") has been notoriously badly handled in bioassessment programs, perhaps because the focus is usually on making informed, timely decisions. Commonly, annual samples of Reference sites are used as independent points in an RCA predictive model, and assessing temporal change in Test sites by seeing if their assessment comes out differently in different years. I'll demonstrate what I think is a better approach to incorporating temporal variation into bioassessments so that trends through time can be discerned and used in sensitive assessments of changes in ecosystem health.
7. **Moving beyond NHST**: In all of the modelling I do in this book I will use a Bayesian fitting approach, which for me captures the science of bioassessment (and maybe other disciplines) well. At its heart, the null-hypothesis-significance-testing (NHST) paradigm is about a binary decision: accept or reject the null hypothesis. It implements that by asking, *What was the chance was of getting the data you got if the null hypothesis was true*? (the infamous "p-value"). Bayesian fitting flips that on its head, asking for a whole bunch of hypotheses, *What is the likelihood of this hypothesis being true given the data we got?* Importantly, all of the other aspects of my rethinking bioassessment can be implemented using NHST if you're more comfortable in that world.

I will illustrate my approach to bioassessment with two datasets: **SimStreams** is a simulated dataset of stream benthic macroinvertebrate communities from 500 streams that vary in Catchment Area and Farming Activity from each of two (imaginary) Ecoregions, and **GLBEAST** is an actual dataset obtained from the Canadian federal Department of Environment & Climate Change (ECCC). It includes benthic macroinvertebrate communities collected from 2005–2019 in nearshore ecosystems in the Canadian Great Lakes (Superior, Huron, Erie, Ontario), as well as Effective Fetch at each site and human activity as quantified with distance to shore, road density, building density, and human population size.

For each of these sample datasets, I will go through the following sequence, with key points to remember at the end of each of Chapters Two to Five as well as the relevant R scripts to carry out the tasks:

- Defining a conceptual model for how the natural and human activity environments cause the biota at each site, describing the natural and human activity environment of a freshwater ecosystem, and defining Reference *versus* Test sites (Chapter Two)
- Modeling the Reference Condition using the relationship between the biota and the natural environment at Reference sites (Chapter Three)
- Bioassessment of Test sites and building diagnosis models to evaluate the effects of specific human activities on ecosystems at Reference and Test sites (Chapter Four)

- Modelling temporal change in freshwater ecosystems with the natural environment at Reference and the human activity environment at Reference and Test sites (Chapter Five; GLBEAST only)

## Chapter One Key Points

- The **logic of bioassessment** is simple – how does what is living in an ecosystem reflect the effects of human activity on that ecosystem?
- In bioassessments carried out over the last 50 years, there has been increased use of **quantitative descriptions of the biota** with diversity and other indices as well as multivariate characterization of communities, primarily using taxonomically categorized abundance, biomass, or presence/absence data.
- The **BACI** (Before-After-Control-Impact) bioassessment design has been seen as a gold standard for more than 40 years, and emphasized a **relative rather than an absolute definition of a healthy ecosystem**.
- The **Reference Condition Approach** (RCA) to bioassessment extended the spatial scale of Control (*aka* Reference) sites in the BACI design to sites less exposed to human activity *versus* Impact (*aka* Test) sites that are more exposed to human activity.

## References

Bailey, R.C., Green, R.H.: Within-basin variation in the shell morphology and growth-rate of a fresh-water mussel. Can. J. Zool. **66**, 1704–1708 (1988)

Baird, D.J., Hajibabaei, M.: Biomonitoring 2.0: a new paradigm in ecosystem assessment made possible by next-generation DNA sequencing. Mol. Ecol. **21**, 2039–2044 (2012)

Baird, D.J., Van den Brink, P.J., Chariton, A.A., Dafforn, K.A., Johnston, E.L.: New diagnostics for multiply stressed marine and freshwater ecosystems: integrating models, ecoinformatics and big data. Mar. Freshw. Res. **67**, 391–392 (2016)

Barbour, M.T., Gerritsen, J., Snyder, B.D., Stribling, J.B.: Rapid Bioassessment Protocols For Use in Streams and Wadeable Rivers: Periphyton, Benthic Macroinvertebrates, and Fish Second Edition. EPA 841-B-99–002, USEPA (1999)

Bonada, N., Prat, N., Resh, V.H., Statzner, B.: Developments in aquatic insect biomonitoring: A comparative analysis of recent approaches. Annu. Rev. Entomol. **51**, 495–523 (2006)

Buss, D.F., Carlisle, D.M., Chon, T.-S. Culp, J., Harding, J.S., Keizer-Vlek, H.E., Robinson, W.A., Strachan, S., Thirion, C., Hughes, R.M.: Stream biomonitoring using macroinvertebrates around the globe: a comparison of large-scale programs. Environmental Monitoring and Assessment **187** (2015)

Costas, N., Pardo, I., Mendez-Fernandez, L., Martinez-Madrid, M., Rodriguez, P.: Sensitivity of macroinvertebrate indicator taxa to metal gradients in mining areas in Northern Spain. Ecol. Ind. **93**, 207–218 (2018)

Downing, J.A.: Aggregation, transformation, and the design of benthos sampling programs. J. Fish. Res. Board Can. **36**, 1454–1463 (1979)

Egerton, F.N.: Part 57: Aspects of Limnology in America, 1930s to about 1990, Led by Hutchinson and Hasler. Bull. Ecol. Soc. Am. **97**, 228–282 (2016)

Egerton, F.N.: History of Ecological Sciences Part 50 Formalizing Limnology, 1870s to 1920s. Bulletin of the Ecological Society of America:131–153 (2014)

Elliott, J.M.: Some methods for the statistical analysis of samples of benthic invertebrates. Freshw. Biol. Assoc. (1977)

Green, R.H.: Sampling design and statistical methods for environmental biologists. Wiley, New York (1979)

Hajibabaei, M., Baird, D.J., Fahner, N.A., Beiko, R., Golding. G.B.: A new way to contemplate Darwin's tangled bank: how DNA barcodes are reconnecting biodiversity science and biomonitoring. Philos. Trans. R. Soc. B-Biol. Sci. **371** (2016)

Hilsenhoff, W.L.: An improved biotic index of organic stream pollution. v **20**, 31–39 (1987)

Hurlbert, S.H.: Pseudoreplication and the design of ecological field experiments. Ecol. Monogr. **54**, 187–211 (1984)

Hynes, H.B.N.: The biology of polluted waters. Liverpool University Press (1966)

Karr, J.R.: Biological Monitoring and Environmental Assessment—A Conceptual-Framework. Environ. Manage. **11**, 249–256 (1987)

Karr, J.R., Larson E.R., Chu, E. W.: Ecological integrity is both real and valuable. Conserv. Sci. Pract. (2021)

Kolkwitz, R., Marrson, M.: Ecology of animal saprobia. Pages 85–95 in Keup, L.E., Ingram, W.M., Mackenthum, K.M. editors. Biology of Water Pollution: A collection of selected papers on stream pollution, waste water, and water treatment. Federal Water Pollution Control Administration, US Department of the Interior (1967a)

Kolkwitz, R., Marrson, M.: Ecology of plant saprobia. Pages 47–51 in Keup, L.E., Ingram, W.M., Mackenthum, K.M., editors. Biology of Water Pollution: A collection of selected papers on stream pollution, waste water, and water treatment. Federal Water Pollution Control Administration, US Department of the Interior (1967b)

Leopold, A., Schwartz, C.W.: A Sand County almanac, and sketches here and there. Oxford University Press, New York (1949)

Maasri, A., Jähnig, S.C., Adamescu, M.C., Adrian, R., Baigun, C., Baird, D.J., Batista-Morales, A., Bonada, N., Brown, L.E., Cai, Q., Campos-Silva, J.V., Clausnitzer, V., Contreras-Macbeath, T., Cooke, S.J., Datry, T., Delacámara, G., De Meester, L., Dijkstra, K.D.B., Do, V.T., Domisch, S., Dudgeon, D., Erös, T., Freitag, H., Freyhof, J., Friedrich, J., Friedrichs-Manthey, M., Geist, J., Gessner, M.O., Goethals, P., Gollock, M., Gordon, C., Grossart, H.P., Gulemvuga, G., Gutiérrez-Fonseca, P.E., Haase, P., Hering, D., Hahn, H.J., Hawkins, C.P., He, F., Heino, J., Hermoso, V., Hogan, Z., Hölker, F., Jeschke, J.M., Jiang, M., Johnson, R.K., Kalinkat, G., Karimov, B.K., Kasangaki, A., Kimirei, I.A., Kohlmann, B., Kuemmerlen, M., Kuiper, J.J., Kupilas, B., Langhans, S.D., Lansdown, R., Leese, F., Magbanua, F.S., Matsuzaki, S.I.S., Monaghan, M.T., Mumladze, L., Muzon, J., Mvogo Ndongo, P.A., Nejstgaard, J.C., Nikitina, O., Ochs, C., Oghenekaro, J.J., Opperman, H.P., Steffen, R., Raghavan, A., Ramírez, B., Rashni, V., Ross-Gillespie, M.J., Samways, R.B., Schäfer, A., Schmidt-Kloiber, O., Seehausen, D.N., Shah, S., Sharma, J., Soininen, N., Sommerwerk, J.D., Stockwell, F., Suhling, R.D., Tachamo Shah, R.E., Tharme, J.H., Thorp, D., Tickner, K., Tockner, J.D., Tonkin, M., Valle, J., Vitule, M., Volk, D., Wang, C.W., Worischka, S.: A global agenda for advancing freshwater biodiversity research. Ecol. Lett. **25**, 255–263 (2022)

MacArthur, R.H., Wilson, E.O.: Equilibrium-theory of insular zoogeography. Evolution **17**, 373–380 (1963)

Metcalfe, J.L., Reynoldson, T.B.: Special section on the use of benthic invertebrates in environmental assessment (Proceedings of A Symposium Held at the 31St Conference on Great-Lakes Research, Mcmaster- University, Hamilton, Ontario)—Preface. Journal of Great Lakes Research **15**:537–537 (1989)

Nichols, S.J., Reynoldson, T.B., Harrison, E.T.: Evaluating AUSRIVAS predictive model performance for detecting simulated eutrophication effects on invertebrate assemblages. Freshwater Science **33**, 1212–1224 (2014)

Norton, S.B., Cormier, S.M., Suter Ii, G.W.: Ecological causal assessment. CRC Press (2015)

Orlofske, J.M., Baird, D.J.: Incorporating continuous trait variation into biomonitoring assessments by measuring and assigning trait values to individuals or taxa. Freshw. Biol. **59**, 477–490 (2014)

Patrick, R.: Structure of diatom communities under varying ecological conditions. Ann. N. Y. Acad. Sci. **108**, 359–360 (1963)

Rawson, D.S.: Some physical and chemical factors in the metabolism of lakes. Pages 9–26 in Moulton, F.R. editor. Problems of lake biology. AAAS, Pennsylvania, USA (1939)

Resh, V.H.: Sampling Variability and Life History Features: Basic Considerations in the Design of Aquatic Insect Life. J.Fish Res.Board Can. **36** (1979)

Rohwer, Y., Marris, E.: Ecosystem integrity is neither real nor valuable. Conserv. Sci. Pract. (2021)

Shannon, C.E.: A mathematical theory of communication—Part One. Bell Syst. Tech. J. **27**, 379–423 (1948)

Somers, K.M., Kilgour, B.W., Munkittrick, K.R., Arciszewski, T.J.: An adaptive environmental effects monitoring framework for assessing the influences of liquid effluents on benthos, water, and sediments in aquatic receiving environments. Integr. Environ. Assess. Manag. **14**, 552–566 (2018)

Stewart-Oaten, A., Murdoch, W.W., Parker, K.R.: Environmental-Impact assessment—pseudoreplication in time. Ecology **67**, 929–940 (1986)

Strachan, S.A., Reynoldson, T.B.: Performance of the standard CABIN method: comparison of BEAST models and error rates to detect simulated degradation from multiple data sets. Freshw. Sci. **33**, 1225–1237 (2014)

Underwood, A.J.: Beyond Baci - the Detection of Environmental Impacts on Populations in the Real, But Variable, World. J. Exp. Mar. Biol. Ecol. **161**, 145–178 (1992)

Walley, W.J., Hawkes, H.A.: A computer-based reappraisal of the Biological Monitoring Working Party scores using data from the 1990 river quality survey of England and Wales. Water Res. **30**, 2086–2094 (1996)

Whorley, S.B., Wehr, J.D.: Multiyear patterns in benthic algal fatty-acid compounds under agricultural stress. Freshwater Science **37**, 534–550 (2018)

Wright, J.F., Moss, D., Armitage, P.D., Furse, M.T.: A preliminary classification of running-water sites in Great Britain based on macro-invertebrate species and the prediction of community type using environmental data. Freshw. Biol. **14**, 221–256 (1984)

Yates, A.G., Bailey, R.C.: Covarying patterns of macroinvertebrate and fish assemblages along natural and human activity gradients: implications for bioassessment. Hydrobiologia **637**, 87–100 (2010b)

Yates, A.G., Culp, J.M., Armanini, D.G., Baird, D.J., Jardine, T.D., Orlofske, J.M.: Enhancing bioassessment approaches: development of a river services assessment framework. Freshw. Sci. **38**, 12–22 (2019)

Yates, A.G., Bailey, R.C.: Selecting objectively defined reference sites for stream bioassessment programs. Environ. Monit. Assess. **170** (2010b)

# Chapter 2
# Building a Conceptual Model and Defining Reference and Test Sites

## The Conceptual Model

Throughout my rethinking of bioassessment, and this book, I have and will continue to emphasize the importance of the conceptual model in getting started with (or rehabilitating) a bioassessment program. Speaking from experience, most programs start with a goal of using the biota to assess the impact of human activity in a given type of ecosystem, but move quickly to thinking about what data are available or what data traditionally get collected as environmental descriptors. The conceptual model is often unstated, and most effort after data collection is expended on how to put together biota and environmental datasets and get straight to the bioassessment analysis.

To illustrate what I think is a better approach, I wanted to start this chapter with conceptual models for each of the two example datasets used in this book. I have kept them really simple to illustrate their utility…you will probably think of other natural and human activity factors to include, particularly for the real GLBEAST dataset, but the point is to start by putting your tentative understanding of the ecosystems you are assessing on paper and refining the conceptual model as it helps you decide what descriptors are important to include or exclude when describing the natural and human activity environments, defining the Reference sites, and building Reference Condition and diagnosis models. The best models, conceptual or otherwise, improve your bioassessment program by making it clear where you're wrong or where the information gaps are.

### *SimStreams*

I consider just two natural and one human activity environmental factors in my SimStreams conceptual model (Fig. 2.1). I think it is reasonable to assume that

 
R. C. Bailey, *Rethinking Bioassessment of Freshwater Ecosystem Health*,
https://doi.org/10.1007/978-3-032-12427-2_2

Catchment Area, Ecoregion, and Farming Activity can all contribute to determining the benthic macroinvertebrates living in one of the SimStreams. Bigger Catchment Areas will have a greater variety of niches and abundance of resources, and provide a bigger target for colonization (MacArthur and Wilson 1963). We would tend to find more diversity and a different composition of invertebrate communities in Valley versus Highland streams for the same reason. And Farming Activity is bound to affect which taxa are present…more of those tolerant of changes in the ecosystem and less of those sensitive to the changes. My conceptual model just formalizes this without presuming interactive or independent effects of these factors, or their magnitude or relative importance. It *does* include the causal effect of Ecoregion on Farming Activity…farming is never done randomly! These sort of confounds need to be recognized, especially when diagnosing why a Test site has different biota than we would expect if it is in Reference Condition (discussed in Chap. 4).

In my conceptual model for SimStreams, I've drawn a yellow line separating out the human activity (farming) environment because when we actually do Reference Condition modeling in Chapter 3, we won't include Farming Activity when we build a model that shows how the biota is affected by the natural environment at Reference sites.

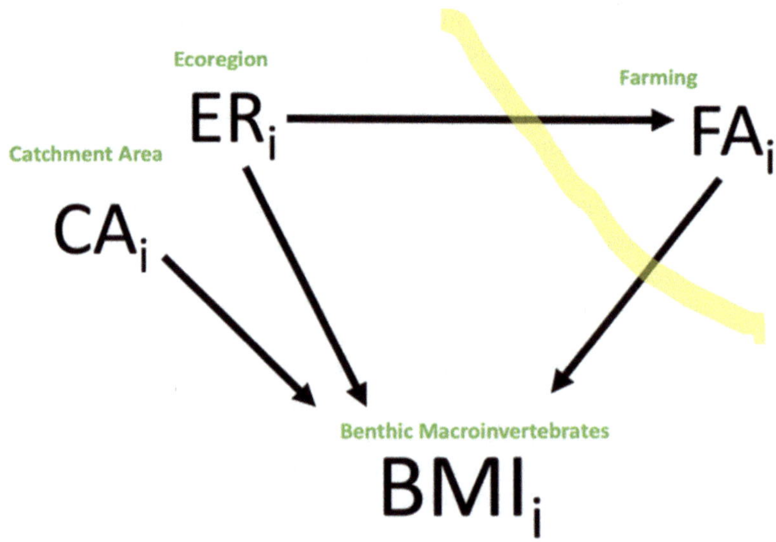


**Fig. 2.1** Conceptual model for benthic invertebrate communities in SimStreams simulated stream ecosystems

## *GLBEAST*

The GLBEAST dataset is from a long-term biomonitoring program carried out by Environment & Climate Change Canada at nearshore Great Lakes sites in Canada. Although the data I use here were collected from 2000–2015 (Fig. 2.2), the program has been around since the early 1990s. It was one of the three example datasets we used in Bailey et al. (2004), and was also used as an example dataset in the methods assessment special issue of Freshwater Science (Bailey et al. 2014).

As with SimStreams, there are just two natural and one human activity environmental factors in my GLBEAST conceptual model (Fig. 2.3). Human Activity, as I'll explain later in this chapter, is a multivariate index based on the first principal component of each site's distance to shore, road and building density, and number of people. Effective fetch (Mason et al. 2018), which is an index of how turbulent a nearshore site is, will cause certain biota to occur and flourish and others to be absent. The same is true for the Great Lake, and the lake also affects human activity; there tend to be more roads, buildings, and people around Lake Ontario than Lake Superior. Just as in the SimStreams conceptual model, I partition the Human Activity from the other, natural environment factors (lake and fetch) so that we can see how unusual Test sites are relative to the biota we expect them to have if they were Reference sites.

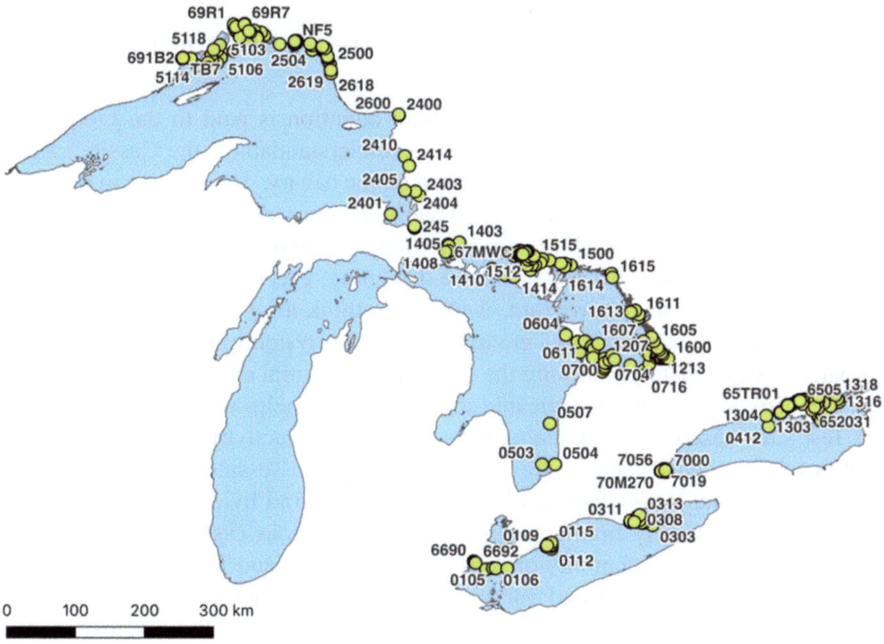


**Fig. 2.2** GLBEAST sampling sites

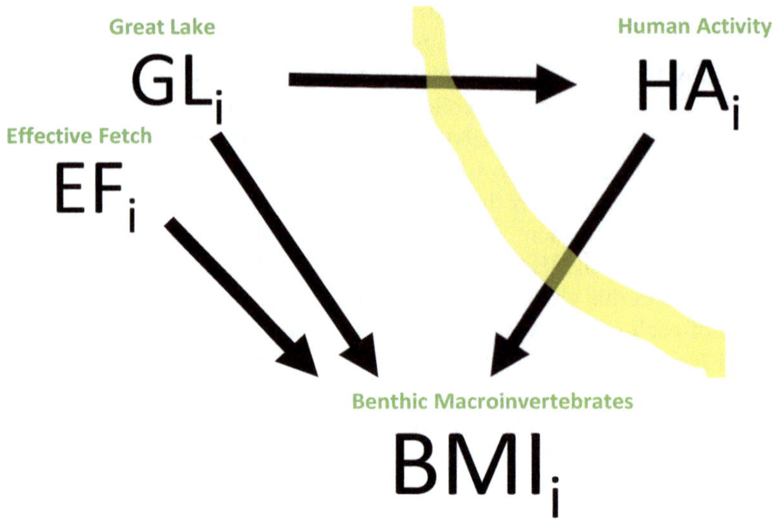


**Fig. 2.3** Conceptual model for benthic invertebrate communities in GLBEAST nearshore Great Lakes ecosystems

## Describing the Natural and Human Activity Environment

In most bioassessment programs, much more attention is paid to the biota in an ecosystem rather than its environment. This is understandable…the classic "canary in the coal mine" bioassessment was focused on the canary. The stressor in the "Test site", deep in a coal mine, was well known (carbon monoxide gas), and the Reference site, a clean cage with ample bird seed in somebody's living room, also was implied if not explicitly defined. But when we are designing, executing, and looking at the results of a bioassessment program, we need to think a lot about both the natural and what I call "human activity" environment of ecosystems. This helps us design a program by adequately sampling the full range of natural environments in a given area (e.g. Yukon River basin wadeable streams). It also helps us to define Reference and Test sites based on each site's exposure to human activity, and then interpret what factors may have caused a Test site to "fail" a bioassessment.

We can't define Reference sites based on what we find living there or how they look…that gets logically circular pretty quickly. Many bioassessment programs have been designed with a dichotomy of "obvious" Reference and Test sites, and careful consideration of the nature and variation of the natural and human activity environments of each site is often a side issue, but I think programs would benefit a lot by making it front and centre. Geographic Information System (GIS) tools have become essential for this process of characterizing and choosing sites in a bioassessment program. In most areas around the world there are great datasets available to characterize both the natural (e.g. bedrock geology), human activity (e.g. road network), and "hybrid" (e.g. climate) environment of each potential or actual (for existing programs) site (Yates et al. 2019b). Most if not all of these tools can be used prior to getting out in the field and sampling, so a stratified random approach can be used to allocate whatever sampling effort is available to characterize Reference ecosystems and their biota. Test sites are usually defined by specific bioassessment program questions or mandates (e.g. are nearshore sites in Hamilton Harbour of Lake Ontario in Reference Condition?), so characterization of the natural environment is needed to predict the biota that *should* be present and the human activity environment can help us understand why a Test ecosystem that is not in Reference Condition failed.

## *The Natural Environment*

One of the best ways to describe the natural environment of aquatic ecosystems actually has nothing to do with the water. **Ecoregions** are defined in various ways to classify terrestrial geographic areas, but for bioassessment I think the best approach is that developed by Robert Bailey (a different and much more famous Robert Bailey!) and colleagues (e.g. Bailey and Ropes 1998). Bailey's ecoregions are based on climate, geology and geography, including biogeography. They are perfect from a bioassessment point of view, since they classify land areas based on where the biota ***should be*** rather than ***where they are***. Because the nature of aquatic ecosystems is always more or less determined by the terrestrial context (Hynes 1975), ecoregions are featured in many conceptual models of bioassessment programs including the SimStreams (simulated) dataset. For my GLBEAST (nearshore Great Lakes) example dataset, rather than using the adjacent terrestrial ecosystem as a descriptor of the natural environment, I used the Great Lake itself that the site is in, which encompasses the effects of surrounding ecoregions as well as limnological differences among the lakes.

Another useful descriptor of the natural environment of aquatic ecosystems is **Catchment Area** (or other variables describing the size and morphology of the ecosystem's drainage basin). The size of a system tends to have an effect on the biota for various reasons going right back to island biogeography theory (MacArthur and Wilson 1963), and Catchment Area is much more temporally stable than descriptors like stream or bankwidth. I include it as one of the descriptors of the natural environment in the SimStreams dataset.

Many other descriptors, at a wide variety of scales, can be used to describe the natural environment. Two key things to keep in mind are:

- ***There is good evidence that the natural environment descriptor causes at least some of the variation in the biota among ecosystems***

This one tends to bother some bioassessment scientists because they are using their judgement as to what to include and what not to include as at least potential predictors in a model. To fulfill this criterion, a descriptor has to vary somewhat from ecosystem to ecosystem and it has to at least theoretically ***cause*** some of the variation in the biota among the ecosystems.

- ***The natural environment descriptor is not affected by human activity***

  For example, *CatchmentArea* is ***not*** affected by human activity in most circumstances, while *% Catchment Area with deciduous forest* **usually is** affected by human activity. The risk if predictors like forest cover or nutrients or sediment particle size are used to describe the environment is that models which predict biota based on an ecosystem's natural environment are at least partially modeling the change in biota with changing human activity (e.g. the effect of removing deciduous forest). The Effective Fetch value I use to describe the natural environment of GLBEAST sites, unlike substrate particle size, is not affected by human activity.

## *The Human Activity Environment*

Bioassessment programs aren't just focused on how the natural environment affects the biota...their primary purpose is to evaluate what effect, if any, the presence or change (increase or decrease) in human activity has on the biota in the ecosystem. Human activity is often categorized as point-source (e.g., effluent from an industrial plant) or diffuse (e.g., cash crop agriculture), and if you add temporal, spatial, and quantitative aspects to the description of human activity....how variable through time is the effluent?....where is the agriculture in relation to the stream network?....how much inorganic fertilizer has been added to the field?...the complexity of describing activity can get pretty overwhelming. In addition, components of this activity will interact with each other and the natural environment to affect which biota are observed in an ecosystem.

The conceptual model for a given bio assessment is important in choosing which descriptors of human activity will be best:

- There is good evidence that the human activity environment descriptor ***causes at least some of the variation in the biota among ecosystems.*** In rural southern Ontario, for example, there is substantial variation in the proportion of Catchment Areas devoted to cash crop agriculture (mostly corn and soybeans). It is reasonable to hypothesize that the proportion of land devoted to cash crops will affect the biota in receiving streams, so this would be a reasonable parameter to include in a model of the effect of human activity on the biota.
- The human activity environment descriptor quantifies the **magnitude and nature** of human activity rather than a **result** of that activity

  It is important that the amount and nature of human activity, rather than its resulting effects on the physical and chemical environment of the ecosystem (e.g.

nitrogen concentration in the water). If we included a descriptor like nitrogen, we would be mixing together variation in nitrogen caused by variation in the natural environment with variation caused by variation in human activity.

## Gradients in the Natural and Human Activity Environments

Although I only use two descriptors of the natural environment with each of my example datasets (one categorical and one quantitative), more often than not there can be several ways to describe it. These can often be integrated into interpretable gradients in the natural environment with an ordination technique like Principal Components Analysis (PCA) as demonstrated in Yates and Bailey (2010b). Similarly, there are often several descriptors of the human activity, and that can be integrated into one or more **Human Activity Gradients** (HAGs) with PCA. PCA of either natural or human activity environment descriptors can be especially useful in including correlated and often confounded descriptors of the natural environment and human activity environment (see McElreath 2020, Chap. 6).

This environmental gradient approach is illustrated with the GLBEAST example dataset, where I integrate distance of the site from shore, road and building density, and population size into one HAG. I greatly prefer this to some sort of statistical "choice" about which descriptors to use and which to discard based on some arbitrary measure of how correlated they are with other descriptors. PCA finds linearly independent axes of environmental variation based on how and how much the different descriptors to the major environmental gradients in the dataset.

## Defining Reference *Versus* Test Sites

The Reference Condition Approach to bioassessment defines healthy ecosystems in a given area by characterizing the biota in those sites that are ***relatively unexposed*** to human activity… *aka* in **Reference Condition** (Bailey et al. 2004). There has been lots of vigorous discussion about how Reference sites are or should be defined. With climate change and other large scale effects of human activity on the environment, that seems like a pretty challenging job. Even without considering these large-scale processes, finding ecosystems in Reference Condition can seem like a daunting task. One of the most thoughtful discussions of this is in Stoddard et al. (2006). The approach I describe here is finding what they call **minimally disturbed sites** to compare to sites that are exposed to a greater amount of human activity. I think it's important to differentiate *exposure* to human activity from the hypothesized or observed *effects* of that activity (cf. Davies and Jackson 2006).

Yates and Bailey (2010) defined a multivariate gradient of agricultural intensity in fourth order stream catchments in southwestern Ontario and then illustrate how to choose the best boundary point between Reference and Test sites in distinct natural areas within the region along such a HAG. They basically choose a classification point where the greatest contrast of human activity between Reference and Test sites but maintaining a decent sample size for building a Reference Condition model. It's a subjective decision, but at least the explicitly defined cards are on the table.

## *SimStreams*

To generate the SimStreams dataset, I used an R script to simulate benthic invertebrate communities from 500 third order streams in two (imaginary) Ecoregions in southern Ontario, "Highland" and "Valley". I'll discuss and model variation in the biota in Chapter Three. The two aspects of the natural environment that were used to characterize each site were:

- **Ecoregion** (ER)—Highland or Valley
- **Catchment Area** of the stream (CA)—from 100 to 1000 ha

These natural environmental descriptors are somewhat correlated, reflecting the reality of pretty much every bioassessment…there tends to be varying amounts of correlation both between and within sets of natural and human activity environmental descriptors. It's reasonable to assume that the Valley Ecoregion tends to have larger Catchment Areas than those in the Highland Ecoregion. If we were designing a program for a given area we could delineate Catchment Areas and then allocate whatever sampling effort we had available across the range of natural environments present. This wouldn't eliminate correlations among descriptors, but it would give us some information about every combination of them that occurred.

Human activity in the SimStreams dataset is described by an index of Farming Activity in the stream's Catchment Area that is scaled from zero to 100. To make the simulation somewhat realistic, I have more intense farming in Valley than Highland Catchment Areas. Bringing descriptions of the natural and human activity environments together in a simple plot (Fig. 2.4) is a great way to communicate the context of the bioassessment program. I think it is absolutely critical to show the range of environments considered in a bioassessment program with simple figures like this prior to the modeling and bioassessment process, which almost always obscure the breadth and potential constraints of the program. Sometimes people skip this simple, descriptive step and immediately jump to variable selection and quite complex models. I think getting to know the data, and maintaining that close relationship, is much more valuable than immediately jumping to sophisticated methods that are somewhat disconnected from the data.

I applied the Yates and Bailey (2010) method of defining Reference *versus* Test sites to the SimStreams dataset with Farming Activity as my HAG. I defined Reference and Test separately for each Ecoregion because of substantially different ranges of Farming Activity in each (Fig. 2.5), reflecting the usual real world situation where human activity is correlated with the natural environment. For Valley Ecoregion

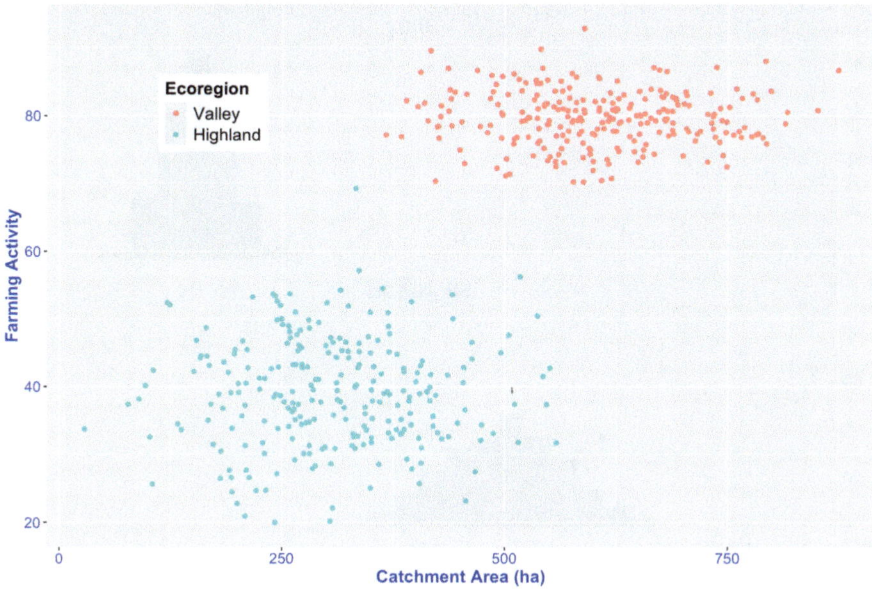


**Fig. 2.4 SimStreams**: Bigger Catchment Areas and greater Farming Activity in Valley *versus* Highland Ecoregions

streams (Fig. 2.6), Reference sites are the 25% with Farming Activity of 75 or less…pretty high for Reference sites! This maximized the difference in Farming Activity between Reference and Test sites while still having a reasonably large number of Reference sites (n = 63) for modeling. For Highland streams, the Reference sites are the lowest 20% on the Human Activity Gradient. These had Farming Activity of 34 or less (Fig. 2.7). There's more of a contrast between Reference and Test sites at lower levels of Farming Activity in both Ecoregions but we would sacrifice sample size to build a relatively robust Reference Condition model. Note that this method is finding the best…or the least exposed…of those sites in each of the Ecoregions, and that the judgement does not in any way consider the biota at the sites.

## *GLBEAST*

As with the SimStreams dataset, I considered just two descriptors of the natural environment that seem likely to at least partially cause the benthic community that was observed at these sites:

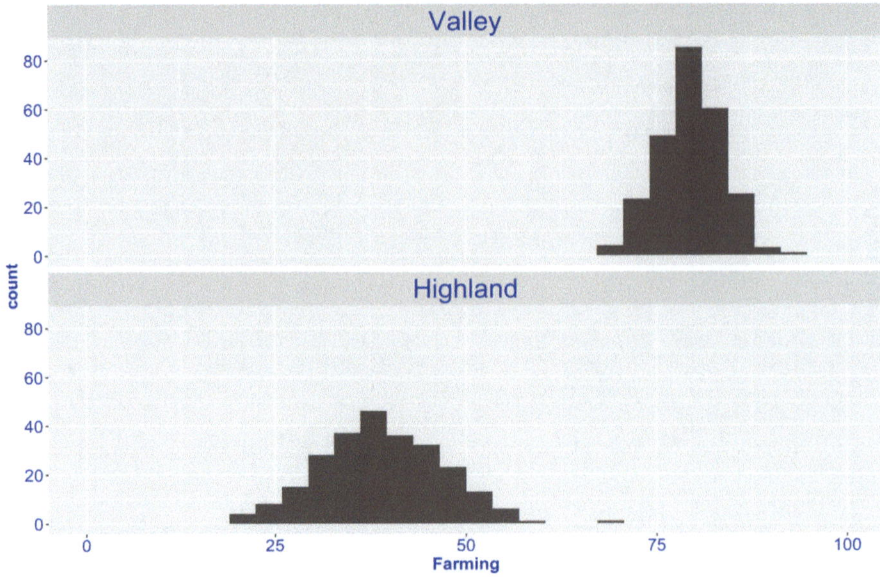


**Fig. 2.5** **SimStreams**: Except for one unusual site in the Highland Ecoregion, there was substantially less Farming Activity in the Highland (lower histogram) *versus* the Valley (upper histogram) Ecoregion

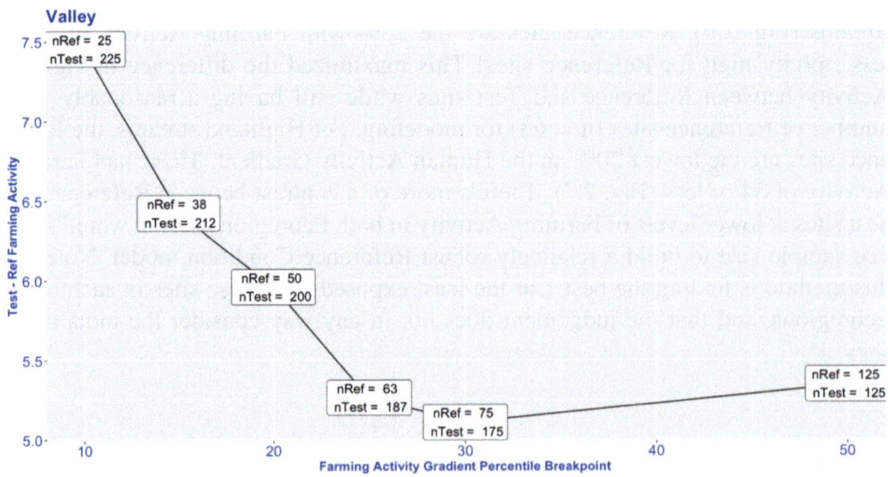


**Fig. 2.6** Classifying Reference and Test sites for SimStreams in the Valley Ecoregion. Sites which have the lowest 25% of Farming activity is decision point that balances differences in exposure to human activity between Reference and Test sites with adequate sample size of Reference (n = 63) and Test (n = 187) sites

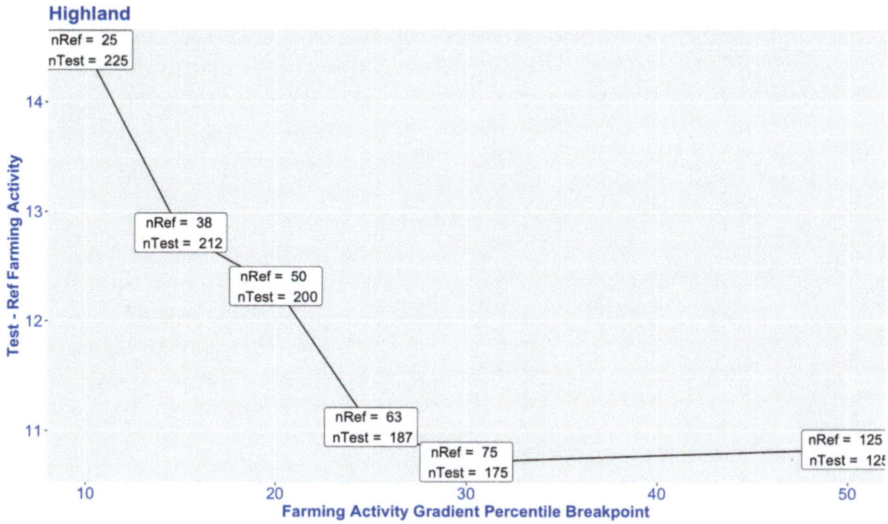


**Fig. 2.7** Classifying Reference and Test sites for SimStreams in the Highland Ecoregion. Sites which have the lowest 20% of Farming activity is decision point that balances differences in exposure to human activity between Reference and Test sites with adequate sample size of Reference (n = 50) and Test (n = 200) sites

- **Lake** (GL)—there are four North American Great Lakes in this dataset: Superior, Huron, Erie, and Ontario. Among many other factors in the natural environment, we might expect to see different benthic invertebrate taxa in each of the four lakes.
- **Effective Fetch** (EF)—the Effective Fetch (Mason et al. 2018) of the ecosystem combines how far the site is from shore in each of 36 compass directions weighted by the proportion of time the wind blew from each direction in a recent year. This has a major impact on turbulence, sediment particle size, and water quality near and within the sediments, and thus the biota that can live in the benthic environment.

Effective fetch varies among the lakes in perhaps unexpected ways. Lake Erie sites have the largest Effective Fetch while Lake Ontario sites have the smallest (Fig. 2.8). Relative to the other lakes, there are more Lake Ontario sites in harbours (e.g. Hamilton Harbour and Presqu'ile Bay) than along open coasts. Keep in mind that this covariation of lake and fetch is much more about the sampling approach of the GLBEAST program during the 2000–2015 period rather than an attempt to capture as much variation as possible in fetch in each of the four lakes. The scope of a bioassessment program is obviously limited to the site's sampled using the program's sampling strategy and available resources.

Note that, as discussed previously, I have avoided descriptors of the natural environment that are affected by human activity (e.g. sediment particle size, sediment and water nutrients) and those not clearly interpretable in a conceptual model (e.g.

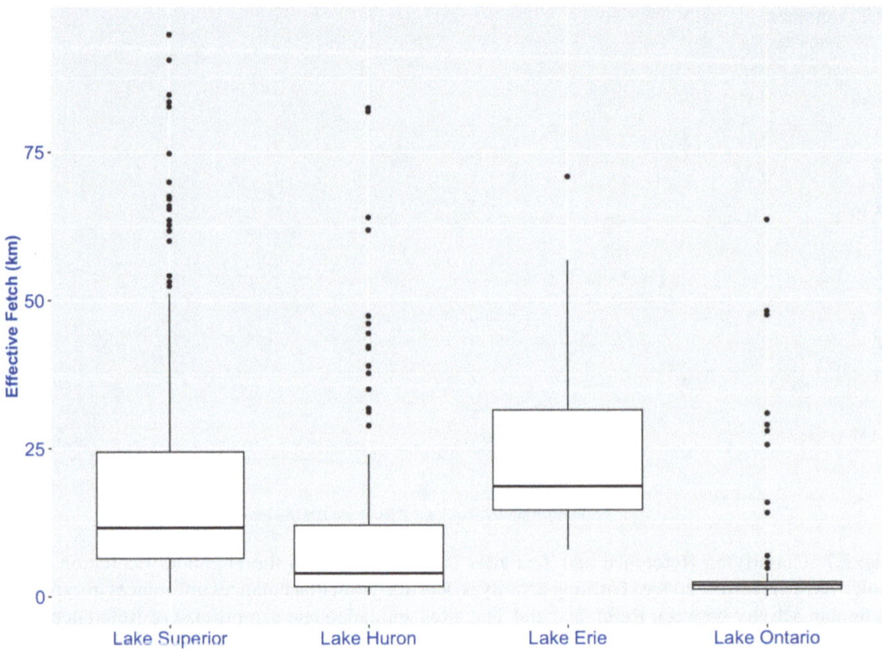


**Fig. 2.8** Effective Fetch (Mason et al. 2018) of GLBEAST sites

latitude, longitude). Bioassessment practitioners often "throw in" latitude and longitude as a descriptor of the natural environment as a recognition that where the site is affects biota. I find it more useful to use ecologically relevant descriptors of location like Great Lake (used in this example) or biogeographic zone or adjacent terrestrial ecoregion.

The GLBEAST dataset is real, so it has all the texture (and challenges) of actual data as each GLBEAST site is a nearshore ecosystem in the Canadian Great Lakes. To illustrate the approach I used, I will zoom in on a few sites in Presqu'ile Bay, Lake Ontario (Fig. 2.9).

I assumed that what is happening on land adjacent to the site is important in describing human activity, so I defined what I call a **terrestrial cookie**, the terrestrial part of a 10 km diameter circle that is centred on the nearest point of land from the site. The distance from the site to the centre of this circle should be a decent, inverse measure of how exposed the site is to the human activity in the terrestrial cookie. I have pretty good GIS layers from publicly available sources describing the number of buildings, road network, and population size (https://geohub.lio.gov.on.ca/), so I ended up with three descriptors of human activity at each GLBEAST site:

- distance from the site to nearest point on shore (m)
- number of buildings per 100 ha in the adjacent land
- km's of roads per 100 ha in the adjacent land
- number of people in the adjacent land

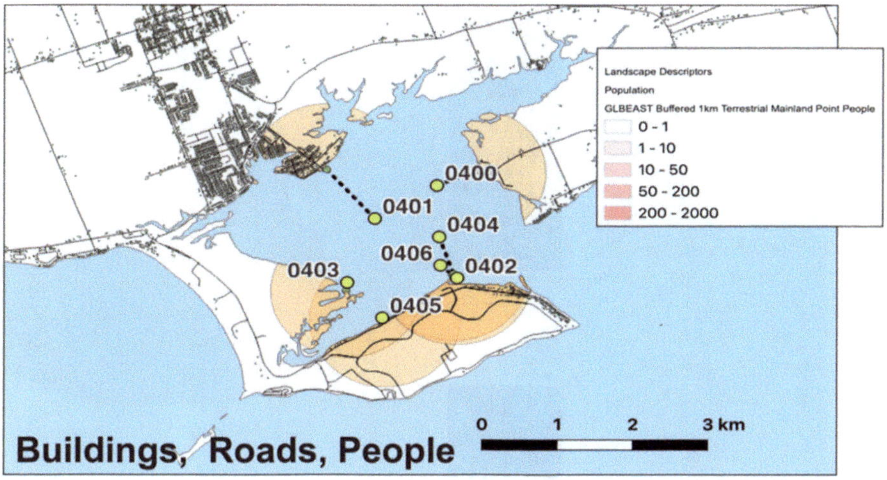


**Fig. 2.9** Buildings, roads, and people in the terrestrial area adjacent to GLBEAST sites in Presqu'ile Bay, Lake Ontario

I used Principal Component Analysis (PCA) to define a Human Activity Gradient for the GLBEAST dataset. First, QGIS was used to define each site's adjacent terrestrial area and then the distance, size, building density, road density, and number of people from available provincial and federal GIS layers. The loadings of the first, major gradient from this PCA (Table 2.1) indicated environmental variation of sites from those that were far from shore with few buildings, roads, and people to sites that were close to shore with lots of buildings, roads, and people (Fig. 2.10).

For the GLBEAST dataset, I defined Reference sites separately in each of the Great Lakes for the same reason I defined Reference sites separately for the Valley and Highland Ecoregions in the SimStreams dataset: there was a difference in the frequency distribution of sites along the HAG in each of the lakes (Fig. 2.11). This resulted in a breakpoint distinguishing Reference from Test sites at the lowest 50%

**Table 2.1** Principal Component Analysis of human activity descriptors at GLBEAST sites. There was just one important gradient (HAG1), mainly defined by distance from the site to the adjacent terrestrial area, building density, road density, and number of people

| | HAG1 | HAG2 | HAG3 |
|---|---|---|---|
| log(Mainland distance) | −0.252 | −0.771 | −0.564 |
| log(Area) | 0.046 | 0.093 | 0.082 |
| log(Perimeter) | 0.012 | 0.049 | 0.061 |
| log(Building density) | 0.699 | −0.512 | 0.360 |
| log(Road density) | 0.272 | −0.156 | 0.152 |
| log(# people) | 0.610 | 0.329 | −0.720 |
| Standard deviation | 3.18 | 1.293 | 1.02 |
| Proportion of variance | 0.76 | 0.13 | 0.08 |
| Cumulative proportion | 0.76 | 0.89 | 0.97 |

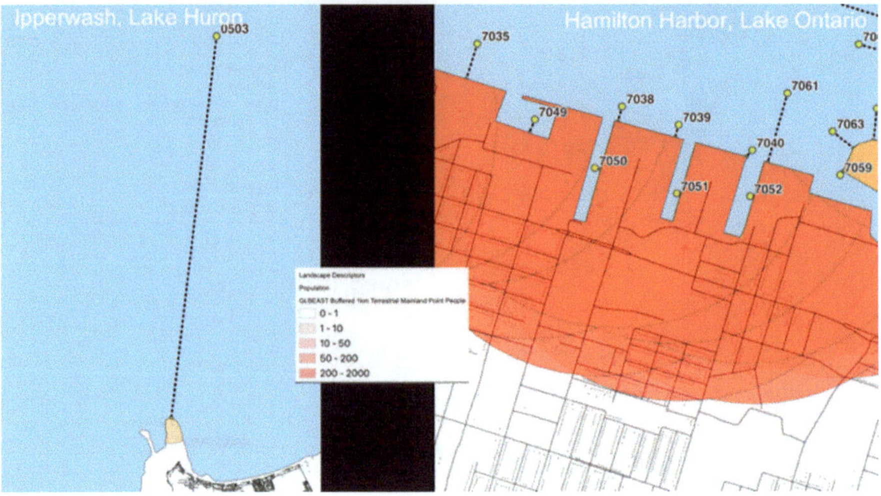


**Fig. 2.10** Extremes along the GLBEAST Human Activity Gradient. Site 503 near Ipperwash in Lake Huron (left) is far from shore with no buildings, roads, or people, while several sites in Hamilton Harbor in Lake Ontario (right) have lots of buildings, roads, and people

of human activity for Superior (Figs. 2.12), Huron (Figs. 2.13), and Erie (Figs. 2.14), and the lowest 30% for Lake Ontario (Fig. 2.15).

In real studies like GLBEAST, it's always useful to map Reference and Test sites, and in this case it reveals the bias and narrowness of scope of the program in all four lakes. Lake Superior sites are largely limited to the north shore, with a few sites in the southeastern part of the lake (Fig. 2.16), while Lake Huron sites are almost all in Georgian Bay (Fig. 2.17), with most Reference sites on the North Shore and mainly Test sites in the south. Sites in Lake Erie were limited to areas in or near Long Point Bay and Rondeau Bay, with a few (Test) sites in west of Point Pelee (Fig. 2.18). Lake Ontario sites were all in either Hamilton Harbour or Presqu'ile Bay (Fig. 2.19).

Such biased sampling of each lake might seem rather foolish in hindsight, but bear in mind the goal of this program was bioassessment of specific areas of interest rather than characterization of the lakes as a whole. Of course, any model we build in Chapter Three and Four to predict the Reference Condition biota we expect at Test sites or try to discern the effects of human activity will only be informed by the sites we have sampled, and this constraint and bias must always be borne in mind. This is obvious in this case, but sometimes less so when practitioners quickly move from assembling the datasets and doing the Reference Condition modeling, the bioassessment, and the diagnosis modeling. It always pays off to check in with the data on maps or with simple statistical descriptions at every stage of executing a bioassessment program.

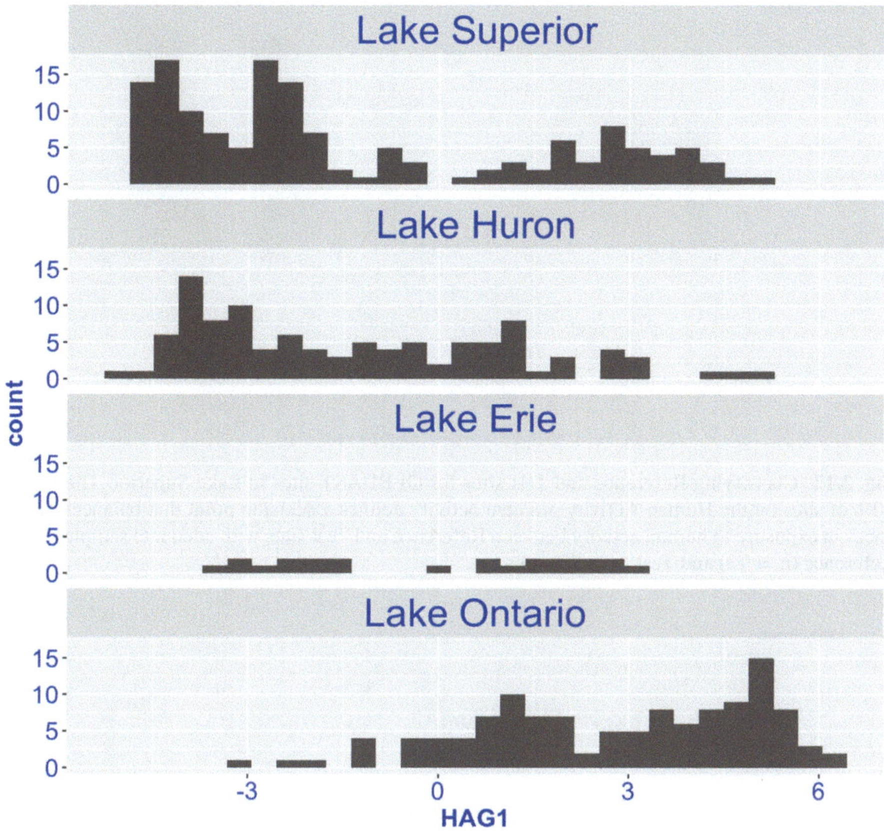


**Fig. 2.11** The varying distribution of sites along the Human Activity Gradient in each of the Great Lakes. Most of the Lake Superior sites are low on the HAG1 axis while most of the Lake Ontario have high Human Activity. Lake Huron and Lake Erie (with a very small sample size) have sites scattered along the gradient

## Chapter Two Key Points

- Start the development of every bioassessment program with a **conceptual model** of how the natural and human activity environments cause variation in the biota from site to site
- Choose descriptors of the natural environment that relate to the conceptual model and are **not affected by human activity**
- Choose descriptors of human activity that relate to the conceptual model and describe the **amount and type** (e.g. proportion of catchment area intensively farmed), rather than the result (e.g. total nitrogen in the water), of human activity
- For both natural and human activity environments, integrate more than one descriptor into one or more indices using a dimensionality reduction method like **Principal Component Analysis**

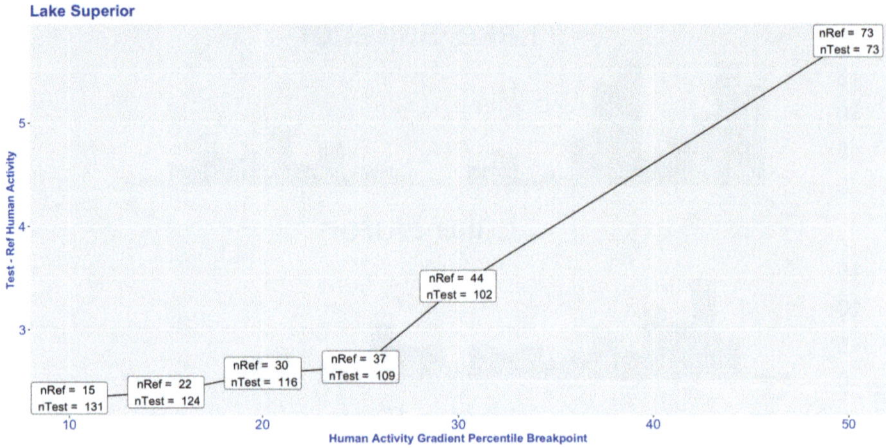


**Fig. 2.12** Classifying Reference and Test sites for GLBEAST sites in Lake Superior. The lowest 50% of sites on the Human Activity gradient activity defines a decision point that balances differences in exposure to human activity between Reference and Test sites with adequate sample size of Reference (n = 73) and Test (n = 73) sites

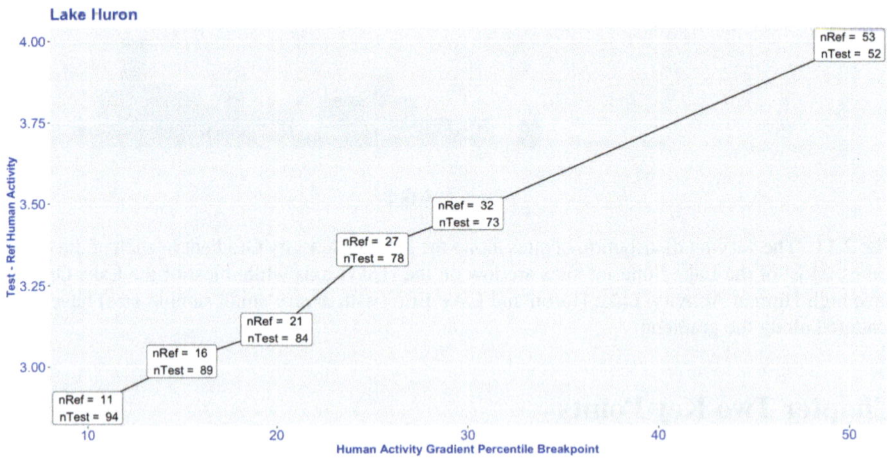


**Fig. 2.13** Classifying Reference and Test sites for GLBEAST sites in Lake Huron. The lowest 50% of sites on the Human Activity gradient activity defines a decision point that balances differences in exposure to human activity between Reference and Test sites with adequate sample size of Reference (n = 53) and Test (n = 52) sites

- Based on the distribution of sites along a gradient of human activity, choose a breakpoint to **distinguish Reference** (lower on the human activity gradient) **from Test** (higher on the human activity gradient) sites. Such a definition may be separate for each level of a categorical variable that describes the natural environment (e.g. Ecoregion, Great Lake)

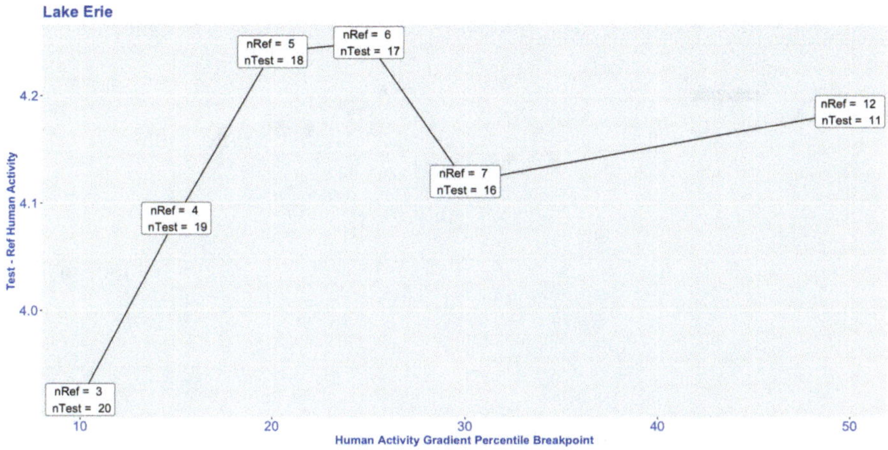


**Fig. 2.14** Classifying Reference and Test sites for GLBEAST sites in Lake Erie. The lowest 50% of sites on the Human Activity gradient activity defines a decision point that balances differences in exposure to human activity between Reference and Test sites with adequate sample size of Reference (n = 12) and Test (n = 11) sites

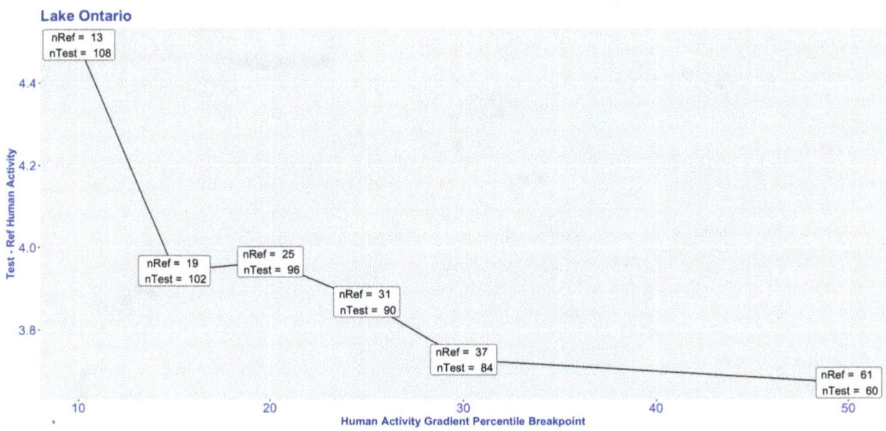


**Fig. 2.15** Classifying Reference and Test sites for GLBEAST sites in Lake Ontario. The lowest 30% of sites on the Human Activity gradient activity defines a decision point that balances differences in exposure to human activity between Reference and Test sites with adequate sample size of Reference (n = 37) and Test (n = 84) sites

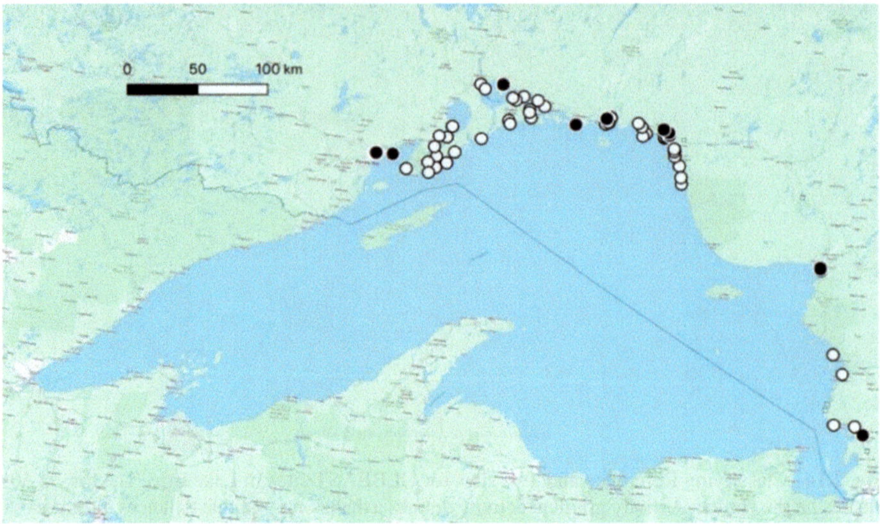


**Fig. 2.16** Reference (n = 73; open circles) and Test (n = 73; closed circles) sites in Lake Superior

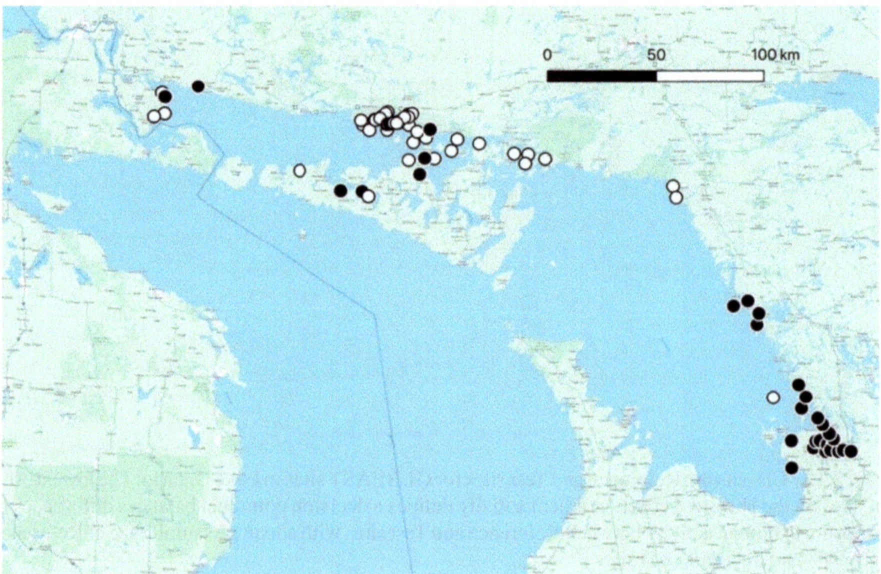


**Fig. 2.17** Reference (n = 53; open circles) and Test (n = 52; closed circles) sites in Lake Huron

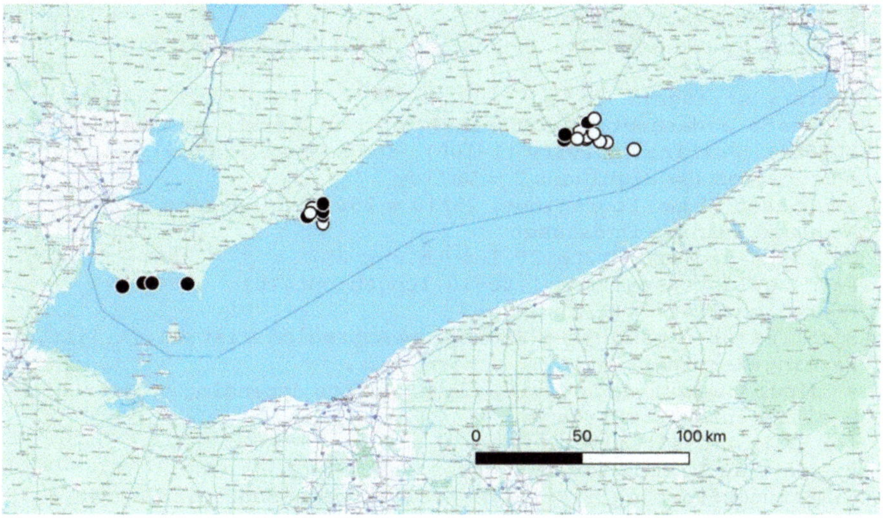


**Fig. 2.18** Reference (n = 12; open circles) and Test (n = 11; closed circles) sites in Lake Erie

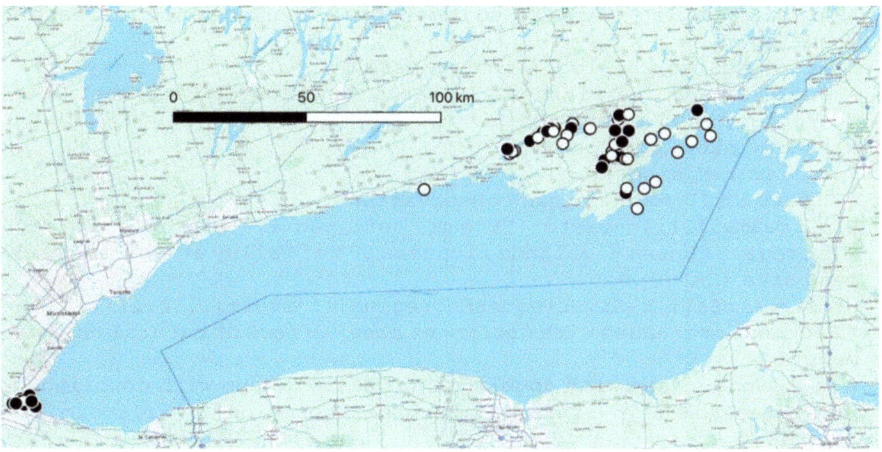


**Fig. 2.19** Reference (n = 37; open circles) and Test (n = 84; closed circles) sites in Lake Ontario

# Chapter Two R Scripts

## *SimStreams*

**SimStreams 0_1 Biota Generator.R – Simulates 500 streams (2 Ecoregions × 25 Catchment Areas × 10 levels of Farming Activity) each with a given species richness.**

```
# Load necessary packages
# [expss] package allows SPSS style labelling of variables
library(expss)
library(stringr)
# Create a dataset [SimStreams] grid of Catchment Area (100 to
1000ha), Farming Activity (1-100), and Ecoregion (1,2) values
# Ecoregion 1 = Highlands 2 = Valley
# For each of two Ecoregions, 25*10 = 250 combinations of Catchment
Area and Farming in dataset
CA.list<-seq(from=1,to=25,length.out=25)
Farming.list<-seq(from=1,to=10,length.out=10)
Ecoregion.list<-c(1,2)
SimStreams<-expand.grid(Ecoregion=Ecoregion.list,CA=CA.list,
Farming=Farming.list)
# Calculate Farming Activity for each site depending on Ecoregion
N_sites<-nrow(SimStreams)
for (i in 1:N_sites) {
# Create an ID code for each observation
SimStreams$Farming[i]<-rnorm(1,SimStreams$Ecoregion[i]*40,8/
SimStreams$Ecoregion[i])
SimStreams$CA[i]<-rnorm(1,SimStreams$Ecoregion[i]*300,100)
}
# Create necessary predictors that are transforms of raw predictors
# In this case log CA is log to base 2 of CA thus one unit increase is
doubling of CA
SimStreams$log_CA<-log2(SimStreams$CA)
# Simulate species richness (S) as a function of Catchment Area,
Ecoregion, and Farming Activity
N_sites<-nrow(SimStreams)
for (i in 1:N_sites) {
# Create an ID code for each observation
SimStreams$ID[i]<-paste('Stream_',str_pad(i,3,pad='0'),sep=")
# Average richness Highlands richness 50% of Valley at a given Catch-
ment area (CA)
SimStreams$a[i]<-SimStreams$Ecoregion[i]*rnorm(1,10,1)
# Richness increases with Catchment area (effect of natural environ-
ment)
# Highlands: 3 spp with doubling CA; Valley: 8 spp with doubling CA
SimStreams$bWS[i]<-ifelse(SimStreams$Ecoregion[i]==1,rnorm(1,3,.3),
rnorm(1,8,.3))
# Farming affects Ecoregions differently
# Highlands: 0.1 spp GAINED per unit Farming increase; Valley: 0.1
spp LOST per unit Farming increase
# This represents a causal simpact model that could be informed by
real data
SimStreams$bFarm[i]<-ifelse(SimStreams$Ecoregion[i]==1,rnorm
(1,0.15,0.01),rnorm(1,-0.3,0.005))
# Summing effects of Natural Environment (Ecoregion, Catchment
Area) and Human Activity (Farming) on Spp
SimStreams$Spp[i]<-round((SimStreams$a[i] +
(SimStreams$bWS[i]*SimStreams$log_CA[i]) + (SimStreams$bFarm[i]
*SimStreams$Farming[i])),digits=0)}
# Label Ecoregions
SimStreams$Ecoregion <- factor(SimStreams$Ecoregion,
```

```
levels = c(2,1),
labels = c("Valley", "Highland"))
# Attach variable labels that will stick with the R dataset for future
plots
SimStreams = apply_labels(SimStreams,
CA = "Catchment Area (ha)",
Farming = "Farming Activity",
Spp = "Species Richness"
)
# Write external R dataset with selected variables in logical order
SimStreams<-SimStreams[,cbind('ID','Ecoregion','CA','log_
CA','Farming','Spp')]
saveRDS(SimStreams,file="./SimStreams/Datasets/
SimStreams.Rds")
```

**SimStreams 0_2 MDS Scores Generator.R – Simulates standardized ordination scores (zNMDS1, zNMDS2) describing community composition in each of the 500 streams.**

```
# Load necessary packages
library(tidyverse)
# Load Simulated streams with Spp data
SimStreams<-readRDS("./SimStreams/Datasets/SimStreams.Rds")
# Simulate NMDS1, NMDS2 as a function of Spp
# Communities with more species tend to have a different composition
N_sites<-nrow(SimStreams)
for (i in 1:N_sites) {
# Summing effects of Natural Environment (Ecoregion, watershed
area) and Human Activity (Farming) on Spp
SimStreams$NMDS1[i]<-(SimStreams$Spp[i] + rnorm(1,0,10)) +
(0.01*SimStreams$Farming[i] + rnorm(1,0,1))
SimStreams$NMDS2[i]<-(SimStreams$Spp[i] + rnorm(1,0,4)) +
((0.01*SimStreams$Farming[i] + rnorm(1,0,1)))
}
# Standardize NMDS Scores
zNMDS1<-scale(SimStreams$NMDS1)
zNMDS2<-scale(SimStreams$NMDS2)
# Just keep the standardized NMDS scores in dataset
SimStreams <- select(SimStreams,!c(NMDS1,NMDS2))
SimStreams <- cbind(SimStreams,zNMDS1,zNMDS2)
# You can write an external R dataset with the labelled dataset
saveRDS(SimStreams,file="./SimStreams/Datasets/
SimStreams.Rds")
```

**SimStreams 2_1 Scatterplot of Farming v CA by Ecoregion.R – Scatter plot of Farming Activity and Catchment Area identifying the Ecoregion of each stream.**

```
# Load necessary package(s)
library(tidyverse)
library(expss)
# Load data for plotting
```

```
SimStreams<-readRDS("./SimStreams/Datasets/SimStreams.Rds")
# Axis labels
qRvar_title<-var_lab(SimStreams$Farming)
qPvar_title<-var_lab(SimStreams$CA)
# Plot Farming Activity versus Catchment Area using Ecoregion as
point color
ggplot(SimStreams,aes(CA,Farming, color=Ecoregion)) +
theme(axis.text = element_text(color = "blue", size = 12)) +
theme(axis.title = element_text(color = "blue", size =
14,face="bold")) +
xlab(qPvar_title) +
ylab(qRvar_title) +
theme(
legend.position = c(0.2, 0.8),
legend.text = element_text(size = 14),
legend.title = element_text(size = 14,face="bold")
) +
geom_point()
```

**SimStreams 2_2 Histogram of HAG by Ecoregion – Frequency distribution of Farming Activity for each Ecoregion.**

```
# Load necessary package(s)
library(tidyverse)
library(expss)
# Load data for plotting
SimStreams<-readRDS("./SimStreams/Datasets/SimStreams.Rds")
# Histogram of Farming for each Ecoregion
ggplot(SimStreams, aes(x=Farming)) +
theme(axis.text = element_text(color = "blue", size = 16)) +
theme(axis.title = element_text(color = "blue", size =
16,face="bold"),
strip.text = element_text(size=25, color = "blue")) +
theme(text=element_text(color = "blue",size=20)) +
xlim(0,100) +
facet_wrap(~ Ecoregion, nrow=2) +
geom_histogram()
```

**SimStreams 2_3 Defining Reference w HAGs Valley.R – For Valley Ecoregion, calculates and plots difference in median Farming Activity between Reference and Test sites for different breakpoints along human activity gradient.**

```
library(GGally)
library(tidyverse)
library(RColorBrewer)
# Load data for plotting
d<-readRDS("./SimStreams/Datasets/SimStreams w NMDS.Rds")
# Pick the subset of observations
# Either use all (Line7) or use a subgroup
plot_Ecoregion<-"Valley"
d<-subset(d,Ecoregion==plot_Ecoregion)
# Plot title
```

```
output_title<-paste(plot_Ecoregion)
hist(d$Farming)
# Get the potential breakpoints along the HAG for defining Reference
v Test
BPPercent<-c(10,15,20,25,30,50)
BPProps<-BPPercent/100
SimStreams_BP<-matrix(quantile(d$Farming, probs = BPProps),6,1)
# Create vectors to hold Status with different breakpoints and
Reference v Test Comparisons
Status10<-c(rep("R",nrow(d)))
Status15<-c(rep("R",nrow(d)))
Status20<-c(rep("R",nrow(d)))
Status25<-c(rep("R",nrow(d)))
Status30<-c(rep("R",nrow(d)))
Status50<-c(rep("R",nrow(d)))
# Create vectors to hold Reference v Test Comparisons for each break-
point
nRef<-c(rep(0,nrow(SimStreams_BP)))
nTest<-c(rep(0,nrow(SimStreams_BP)))
mHAG_Ref<-c(rep(0,nrow(SimStreams_BP)))
mHAG_Test<-c(rep(0,nrow(SimStreams_BP)))
diff<-mHAG_Test-mHAG_Ref
# Loop over sites to define Reference v Test for each possible break-
point
for(Site_row in 1:nrow(d)) {
Status10[Site_row]<-ifelse(d$Farming[Site_row]>SimStreams_
BP[1],"T","R")
Status15[Site_row]<-ifelse(d$Farming[Site_row]>SimStreams_
BP[2],"T","R")
Status20[Site_row]<-ifelse(d$Farming[Site_row]>SimStreams_
BP[3],"T","R")
Status25[Site_row]<-ifelse(d$Farming[Site_row]>SimStreams_
BP[4],"T","R")
Status30[Site_row]<-ifelse(d$Farming[Site_row]>SimStreams_
BP[5],"T","R")
Status50[Site_row]<-ifelse(d$Farming[Site_row]>SimStreams_
BP[6],"T","R")
}
# Append Status for each breakpoint to HAG data for each site and save
d<-cbind(d,Status10,Status15,Status20,Status25,Status30,Status50)
# Also add Status appropriate for Ecoregion based on Yates analysis
d$Status<-Status25
fnStatus<-paste("./SimStreams/Datasets/SimStreams     w     HAGs
Status",plot_Ecoregion,".Rds")
saveRDS(d,file=fnStatus)
#Collect up medians and sample sizes from Reference and Test groups
for each breakpoint
# Medians
BP10<-tapply(d$Farming,d$Status10,median)
BP15<-tapply(d$Farming,d$Status15,median)
BP20<-tapply(d$Farming,d$Status20,median)
BP25<-tapply(d$Farming,d$Status25,median)
BP30<-tapply(d$Farming,d$Status30,median)
BP50<-tapply(d$Farming,d$Status50,median)
```

```
# Sample sizes
N10<-tapply(d$Farming,d$Status10,length)
N15<-tapply(d$Farming,d$Status15,length)
N20<-tapply(d$Farming,d$Status20,length)
N25<-tapply(d$Farming,d$Status25,length)
N30<-tapply(d$Farming,d$Status30,length)
N50<-tapply(d$Farming,d$Status50,length)
# Now build the table that will enable the best decision about break-
point
# 10%
nRef[1]<-N10[1]
nTest[1]<-N10[2]
mHAG_Ref[1]<-BP10[1]
mHAG_Test[1]<-BP10[2]
diff[1]<-mHAG_Test[1]-mHAG_Ref[1]
# 15%
nRef[2]<-N15[1]
nTest[2]<-N15[2]
mHAG_Ref[2]<-BP15[1]
mHAG_Test[2]<-BP15[2]
diff[2]<-mHAG_Test[2]-mHAG_Ref[2]
# 20%
nRef[3]<-N20[1]
nTest[3]<-N20[2]
mHAG_Ref[3]<-BP20[1]
mHAG_Test[3]<-BP20[2]
diff[3]<-mHAG_Test[3]-mHAG_Ref[3]
# 25%
nRef[4]<-N25[1]
nTest[4]<-N25[2]
mHAG_Ref[4]<-BP25[1]
mHAG_Test[4]<-BP25[2]
diff[4]<-mHAG_Test[4]-mHAG_Ref[4]
# 30%
nRef[5]<-N30[1]
nTest[5]<-N30[2]
mHAG_Ref[5]<-BP30[1]
mHAG_Test[5]<-BP30[2]
diff[5]<-mHAG_Test[5]-mHAG_Ref[5]
# 50%
nRef[6]<-N50[1]
nTest[6]<-N50[2]
mHAG_Ref[6]<-BP50[1]
mHAG_Test[6]<-BP50[2]
diff[6]<-mHAG_Test[6]-mHAG_Ref[6]
# Collect info for each breakpoint and save
SimStreams_BP<-data.frame(cbind(BPPercent,BPProps,    nRef,mHAG_
Ref,nTest,mHAG_Test,diff))
fnBP<-paste("./SimStreams/Datasets/SimStreams        BP",plot_
Ecoregion,".Rds")
saveRDS(SimStreams_BP,file=fnBP)
# Plot the difference between median HAG for Reference v Test against
breakpoint percentage
```

```
BPnlabels<-paste("nRef = ",as.character(nRef),"\n nTest =
",as.character(nTest))
d2<-cbind(SimStreams_BP,BPnlabels)
ggplot(d2, aes(BPPercent,diff,label=BPnlabels)) +
theme(axis.text = element_text(color = "blue", size = 16)) +
theme(axis.title = element_text(color = "blue", size =
14,face="bold")) +
ggtitle(output_title) +
theme(title = element_text(color = "blue", size = 16,face="bold"))
+
xlim(10,50) +
xlab("Human Activity Gradient Percentile Breakpoint") +
ylab("Test - Ref Human Activity") +
geom_point() +
geom_line() +
geom_label(size=5)
```

**SimStreams 2_3 Defining Reference w HAGs Highland.R – For Highland Ecoregion, calculates and plots difference in median Farming Activity between Reference and Test sites for different breakpoints along human activity gradient.**

```
library(GGally)
library(tidyverse)
library(RColorBrewer)
# Load data for plotting
d<-readRDS("./SimStreams/Datasets/SimStreams w NMDS.Rds")
# Pick the subset of observations
# Either use all (Line7) or use a subgroup
plot_Ecoregion<-"Highland"
d<-subset(d,Ecoregion==plot_Ecoregion)
# Plot title
output_title<-paste(plot_Ecoregion)
hist(d$Farming)
# Get the potential breakpoints along the HAG for defining Reference
v Test
BPPercent<-c(10,15,20,25,30,50)
BPProps<-BPPercent/100
SimStreams_BP<-matrix(quantile(d$Farming, probs = BPProps),6,1)
# Create vectors to hold Status with different breakpoints and
Reference v Test Comparisons
Status10<-c(rep("R",nrow(d)))
Status15<-c(rep("R",nrow(d)))
Status20<-c(rep("R",nrow(d)))
Status25<-c(rep("R",nrow(d)))
Status30<-c(rep("R",nrow(d)))
Status50<-c(rep("R",nrow(d)))
# Create vectors to hold Reference v Test Comparisons for each break-
point
nRef<-c(rep(0,nrow(SimStreams_BP)))
nTest<-c(rep(0,nrow(SimStreams_BP)))
mHAG_Ref<-c(rep(0,nrow(SimStreams_BP)))
mHAG_Test<-c(rep(0,nrow(SimStreams_BP)))
diff<-mHAG_Test-mHAG_Ref
```

```
# Loop over sites to define Reference v Test for each possible break-
point
for(Site_row in 1:nrow(d)) {
Status10[Site_row]<-ifelse(d$Farming[Site_row]>SimStreams_
BP[1],"T","R")
Status15[Site_row]<-ifelse(d$Farming[Site_row]>SimStreams_
BP[2],"T","R")
Status20[Site_row]<-ifelse(d$Farming[Site_row]>SimStreams_
BP[3],"T","R")
Status25[Site_row]<-ifelse(d$Farming[Site_row]>SimStreams_
BP[4],"T","R")
Status30[Site_row]<-ifelse(d$Farming[Site_row]>SimStreams_
BP[5],"T","R")
Status50[Site_row]<-ifelse(d$Farming[Site_row]>SimStreams_
BP[6],"T","R")
}
# Append Status for each breakpoint to HAG data for each site and save
d<-cbind(d,Status10,Status15,Status20,Status25,Status30,Status50)
# Also add Status appropriate for Ecoregion based on Yates analysis
d$Status<-Status25
fnStatus<-paste("./SimStreams/Datasets/SimStreams    w    HAGs
Status",plot_Ecoregion,".Rds")
saveRDS(d,file=fnStatus)
#Collect up medians and sample sizes from Reference and Test groups
for each breakpoint
# Medians
BP10<-tapply(d$Farming,d$Status10,median)
BP15<-tapply(d$Farming,d$Status15,median)
BP20<-tapply(d$Farming,d$Status20,median)
BP25<-tapply(d$Farming,d$Status25,median)
BP30<-tapply(d$Farming,d$Status30,median)
BP50<-tapply(d$Farming,d$Status50,median)
# Sample sizes
N10<-tapply(d$Farming,d$Status10,length)
N15<-tapply(d$Farming,d$Status15,length)
N20<-tapply(d$Farming,d$Status20,length)
N25<-tapply(d$Farming,d$Status25,length)
N30<-tapply(d$Farming,d$Status30,length)
N50<-tapply(d$Farming,d$Status50,length)
# Now build the table that will enable the best decision about break-
point
# 10%
nRef[1]<-N10[1]
nTest[1]<-N10[2]
mHAG_Ref[1]<-BP10[1]
mHAG_Test[1]<-BP10[2]
diff[1]<-mHAG_Test[1]-mHAG_Ref[1]
# 15%
nRef[2]<-N15[1]
nTest[2]<-N15[2]
mHAG_Ref[2]<-BP15[1]
mHAG_Test[2]<-BP15[2]
diff[2]<-mHAG_Test[2]-mHAG_Ref[2]
# 20%
```

```
nRef[3]<-N20[1]
nTest[3]<-N20[2]
mHAG_Ref[3]<-BP20[1]
mHAG_Test[3]<-BP20[2]
diff[3]<-mHAG_Test[3]-mHAG_Ref[3]
# 25%
nRef[4]<-N25[1]
nTest[4]<-N25[2]
mHAG_Ref[4]<-BP25[1]
mHAG_Test[4]<-BP25[2]
diff[4]<-mHAG_Test[4]-mHAG_Ref[4]
# 30%
nRef[5]<-N30[1]
nTest[5]<-N30[2]
mHAG_Ref[5]<-BP30[1]
mHAG_Test[5]<-BP30[2]
diff[5]<-mHAG_Test[5]-mHAG_Ref[5]
# 50%
nRef[6]<-N50[1]
nTest[6]<-N50[2]
mHAG_Ref[6]<-BP50[1]
mHAG_Test[6]<-BP50[2]
diff[6]<-mHAG_Test[6]-mHAG_Ref[6]
# Collect info for each breakpoint and save
SimStreams_BP<-data.frame(cbind(BPPercent,BPProps, nRef,mHAG_
Ref,nTest,mHAG_Test,diff))
fnBP<-paste("./SimStreams/Datasets/SimStreams BP",plot_
Ecoregion,".Rds")
saveRDS(SimStreams_BP,file=fnBP)
# Plot the difference between median HAG for Reference v Test against
breakpoint percentage
BPnlabels<-paste("nRef = ",as.character(nRef),"\n nTest =
",as.character(nTest))
d2<-cbind(SimStreams_BP,BPnlabels)
ggplot(d2, aes(BPPercent,diff,label=BPnlabels)) +
theme(axis.text = element_text(color = "blue", size = 16)) +
theme(axis.title = element_text(color = "blue", size =
16,face="bold")) +
ggtitle(output_title) +
theme(title = element_text(color = "blue", size = 16,face="bold"))
+
xlim(10,50) +
xlab("Human Activity Gradient Percentile Breakpoint") +
ylab("Test - Ref Human Activity") +
geom_point() +
geom_line() +
geom_label(size=5)
```

**SimStreams 2_4 Merge Ecoregions w Status.R – Merges streams each Ecoregion with their status as defined in SimStreams 2_3.**

```
library(GGally)
library(tidyverse)
```

```
d1<-readRDS("./SimStreams/Datasets/SimStreams w HAGs Status
Valley.Rds")
d2<-readRDS("./SimStreams/Datasets/SimStreams w HAGs Status High-
land.Rds")
d<-rbind(d1,d2)
saveRDS(d,file="./SimStreams/Datasets/SimStreams w Status.Rds")
```

## *GLBEAST*

### GLBEAST 0_1 Reading and Saving Benthos NatEnv and Human Activity.R – Reads and match merges nearshore Great Lakes environment and biota data.

```
# Load necessary package(s)
# readxl and writexl packages handle XL spreadsheets without Java
library(readxl)
library(writexl)
# Load in CABIN export data from Excel workbook spreadsheet
dNE<-read_excel("./GLBEAST Datasets/GLBEAST Site Info and
Benthos.xlsx",sheet=1)
# Pick out just the columns you want for further analysis
dNE<-cbind(dNE[,1],dNE[,3:4],dNE[,7],dNE[,9])
dBI<-read_excel("./GLBEAST Datasets/GLBEAST Site Info and
Benthos.xlsx",sheet=2)
dHA<-read_excel("./GLBEAST Datasets/GLBEAST Human
Activity.xlsx")
dHA<-cbind(dHA[,2],dHA[,5:10],dHA[,12])
# Merge datasets by site into one dataset containing NE, HA, BI
dENV<-merge(dNE,dHA,by="Site")
dALL<-merge(dENV,dBI,by="Site")
# Write external R dataset
saveRDS(dALL,file="./GLBEAST Datasets/GLBEAST NE HA Benthos.Rds")
```

### GLBEAST 0_2 GLBEAST NMDS Ordination of PA Benthos.R – NMDS ordination of all sites based on biota presence/absence data.

```
# Make sure vegan loaded
library(vegan)
library(car)
library(tidyverse)
library(ggplot2)
library(RColorBrewer)
# Load and filter data
GLBEAST<-readRDS("./GLBEAST/Datasets/GLBEAST.Rds")
# Choose subset of observations if necessary
# GLBEAST<-filter(GLBEAST, SixGroup=="4A")
# GLBEAST<-subset(GLBEAST[GLBEAST$Site != "1601",])
# Define set of variables to use for distance matrix - choose either
quantitative or PA of taxa
```

```
# PA taxa
# Col_1st_Tax is first taxon. Col_Last_Tax is last taxon
Col_1st_Tax<-169 # First column of PA taxa
# Col_1st_Tax<-18 # First column of quantitative taxa
Col_Last_Tax<-314 # Last column of PA taxa
# Col_Last_Tax<-158 # Last column of quantitative taxa
ord_GLBEAST<-cbind(GLBEAST[,Col_1st_Tax:Col_Last_Tax])
# Create the distance matrix
ord_GLBEASTDist<-vegdist(ord_GLBEAST,method="bray")
# Do the Non-metric Multi-Dimensional Scaling ordination
ord_GLBEASTNMDS<-monoMDS(ord_GLBEASTDist)
ord_GLBEASTNMDS
qstress<-signif(ord_GLBEASTNMDS$stress, digits=3)
astress<-paste("Stress = ", as.character(qstress))
d<-GLBEAST[,1:10]
# Standardize NMDS Scores
# PA NMDS
d$z_PA_MDS1<-scale(ord_GLBEASTNMDS$points[,1])
d$z_PA_MDS2<-scale(ord_GLBEASTNMDS$points[,2])
# Quantitative NMDS
# z_Q_MDS1<-scale(ord_GLBEASTNMDS$points[,1])
# z_Q_MDS2<-scale(ord_GLBEASTNMDS$points[,2])
ggplot(data=d,aes(x=z_PA_MDS1, y=z_PA_MDS2)) +
scale_colour_manual(values=c(5,1)) +
geom_point(size=2) +
geom_text(aes(label=ifelse(((z_PA_MDS2>2) | (z_PA_MDS2< -2) | (z_
PA_MDS1< -2) | (z_PA_MDS1> 2)),GreatLake,' '),hjust=1,vjust=-
1),size=3,color="black") +
labs(caption = astress,
x = "z_PA_MDS1", y = "z_PA_MDS2")
```

**GLBEAST 2_1 Boxplot of Fetch by Lake.R – Does a boxplot of Effective Fetch for each of the Great Lakes sampled.**

```
# Load necessary package(s).
library(tidyverse)
library(expss)
# Load data for plotting
GLBEAST_HA<-readRDS("./GLBEAST/Datasets/GLBEAST Env.Rds")
# choice of quantitative variable
qvar_title<-var_lab(GLBEAST_HA$Fetch)
# Boxplots of qvar by Great Lake
ggplot(GLBEAST_HA) +
theme(axis.text = element_text(color = "blue", size = 15)) +
theme(axis.title = element_text(color = "blue", size =
14,face="bold")) +
geom_boxplot(aes(GreatLake,Fetch), position = 'identity') +
xlab(" ") +
# facet_wrap(~ Status) +
ylab("Effective Fetch (km)")
```

**GLBEAST 2_2 PCA Human Activity.R – Does a Principal Component Analysis of the correlation matrix of human activity descriptors from all sites.**

```
library(GGally)
library(tidyverse)
library(RColorBrewer)
# Plot title
glake<-"All Lakes"
output_title<-paste("GLBEAST Human Activity:",glake)
# Load data for plotting
GLBEAST_HA<-readRDS("./GLBEAST/Datasets/GLBEAST Env.Rds")
# Pick the subset of observations
# Either use all (Line14) or use a subgroup (Line16)
d<-GLBEAST_HA
# Exclude following line if PCA on whole dataset
# d<-subset(d,GreatLake==glake)
# Log transformations
d$log_area<-log(d$Area_ha)
d$log_perim<-log(d$Perim_km)
d$log_mainland<-log(d$Mainland_m)
d$log_bldgs<-log(d$nBldgs+1)
d$log_roads<-log(d$Road_km+1)
d$log_people21<-log(d$nPeople21+1)
# Do the PCA on named variables, either correlation (if non-commensurate) or covariance (if commensurate) matrix
PCA.result<-princomp(~d$log_area + d$log_perim + d$log_mainland + d$log_bldgs + d$log_roads + d$log_people21,cor=FALSE,scores=TRUE)
# PCA Results with Scree plot
plot(PCA.result)
summary(PCA.result)
print(PCA.result$loadings,cutoff=0)
# New dataset with scores
HAG_PC1<-PCA.result$scores[,1]
HAG_PC2<-PCA.result$scores[,2]
GLBEAST_HAGs<-cbind(d,HAG_PC1,HAG_PC2)
saveRDS(GLBEAST_HAGs,file="./GLBEAST/Datasets/GLBEAST Env w HAGs v2.Rds")
# Subset observations if desired
GLBEAST_HAGs<-subset(GLBEAST_HAGs,GreatLake == 'Lake Ontario')
# Histogram of main HAG (PC1)
ggplot(GLBEAST_HAGs, aes(HAG_PC1)) +
theme(text = element_text(size = 15)) +
geom_histogram(bins=10,fill="white",col=I("blue"))
# Histogram of main HAG (PC1) by Lake
ggplot(data=GLBEAST_HAGs,aes(HAG_PC1)) +
theme(text = element_text(size = 30, face='bold')) +
geom_histogram(bins=10,fill="white",col=I("blue")) +
facet_grid(~ GreatLake)
```

**GLBEAST 2_3 HAG distn by Great Lake.R – Frequency distribution of Human Activity PC1 score for each of the four Great Lakes.**

```
library(GGally)
library(tidyverse)
library(RColorBrewer)
```

```
library(expss)
library(reporter)
# Plot title
output_title<-paste("GLBEAST Human Activity")
output_subtitle<-("Comparing Definitions of Reference")
# Load data
d<-readRDS("./GLBEAST/Datasets/GLBEAST w HAGs Status Yates.Rds")
# Change histogram plot line colors by groups
ggplot(d, aes(x=HAG_PC1)) +
theme(axis.text = element_text(color = "blue", size = 16)) +
theme(axis.title = element_text(color = "blue", size = 16,face="bold"),
strip.text = element_text(size=25, color = "blue")) +
theme(text=element_text(color = "blue",size=20)) +
facet_wrap(~ GreatLake, nrow=4) +
geom_histogram()
```

**GLBEAST 2_4 Defining Reference w HAGs Lake Superior.R – For Lake Superior, calculates and plots difference in median Human Activity between Reference and Test sites for different breakpoints along human activity gradient.**

```
library(GGally)
library(tidyverse)
library(RColorBrewer)
# Load data for plotting
d<-readRDS("./GLBEAST/Datasets/GLBEAST Env w HAGs.Rds")
# Plot title
glake<-"Lake Superior"
output_title<-paste(glake)
output_subtitle<-("Defining Reference v Test")
# Pick the subset of observations
# Either use all (Line7) or use a subgroup
d<-subset(d,GreatLake==glake)
hist(d$HAG_PC1)
# Get the potential breakpoints along the HAG for defining Reference v Test
BPPercent<-c(10,15,20,25,30,50)
BPProps<-BPPercent/100
GLBEAST_BP<-matrix(quantile(d$HAG_PC1, probs = BPProps),6,1)
# Create vectors to hold Status with different breakpoints and Reference v Test Comparisons
Status10<-c(rep("R",nrow(d)))
Status15<-c(rep("R",nrow(d)))
Status20<-c(rep("R",nrow(d)))
Status25<-c(rep("R",nrow(d)))
Status30<-c(rep("R",nrow(d)))
Status50<-c(rep("R",nrow(d)))
# Create vectors to hold Reference v Test Comparisons for each break-point
nRef<-c(rep(0,nrow(GLBEAST_BP)))
nTest<-c(rep(0,nrow(GLBEAST_BP)))
mHAG_Ref<-c(rep(0,nrow(GLBEAST_BP)))
mHAG_Test<-c(rep(0,nrow(GLBEAST_BP)))
```

```
diff<-mHAG_Test-mHAG_Ref
# Loop over sites to define Reference v Test for each possible break-
point
for(Site_row in 1:nrow(d)) {
Status10[Site_row]<-ifelse(d$HAG_PC1[Site_row]>GLBEAST_
BP[1],"T","R")
Status15[Site_row]<-ifelse(d$HAG_PC1[Site_row]>GLBEAST_
BP[2],"T","R")
Status20[Site_row]<-ifelse(d$HAG_PC1[Site_row]>GLBEAST_
BP[3],"T","R")
Status25[Site_row]<-ifelse(d$HAG_PC1[Site_row]>GLBEAST_
BP[4],"T","R")
Status30[Site_row]<-ifelse(d$HAG_PC1[Site_row]>GLBEAST_
BP[5],"T","R")
Status50[Site_row]<-ifelse(d$HAG_PC1[Site_row]>GLBEAST_
BP[6],"T","R")
}
# Append Status for each breakpoint to HAG data for each site and save
d<-cbind(d,Status10,Status15,Status20,Status25,Status30,Status50)
fnStatus<-paste("./GLBEAST/Datasets/GLBEAST w HAGs
Status",glake,".Rds")
saveRDS(d,file=fnStatus)
#Collect up medians and sample sizes from Reference and Test groups
for each breakpoint
# Medians
BP10<-tapply(d$HAG_PC1,d$Status10,median)
BP15<-tapply(d$HAG_PC1,d$Status15,median)
BP20<-tapply(d$HAG_PC1,d$Status20,median)
BP25<-tapply(d$HAG_PC1,d$Status25,median)
BP30<-tapply(d$HAG_PC1,d$Status30,median)
BP50<-tapply(d$HAG_PC1,d$Status50,median)
# Sample sizes
N10<-tapply(d$HAG_PC1,d$Status10,length)
N15<-tapply(d$HAG_PC1,d$Status15,length)
N20<-tapply(d$HAG_PC1,d$Status20,length)
N25<-tapply(d$HAG_PC1,d$Status25,length)
N30<-tapply(d$HAG_PC1,d$Status30,length)
N50<-tapply(d$HAG_PC1,d$Status50,length)
# Now build the table that will enable the best decision about break-
point
# 10%
nRef[1]<-N10[1]
nTest[1]<-N10[2]
mHAG_Ref[1]<-BP10[1]
mHAG_Test[1]<-BP10[2]
diff[1]<-mHAG_Test[1]-mHAG_Ref[1]
# 15%
nRef[2]<-N15[1]
nTest[2]<-N15[2]
mHAG_Ref[2]<-BP15[1]
mHAG_Test[2]<-BP15[2]
diff[2]<-mHAG_Test[2]-mHAG_Ref[2]
# 20%
nRef[3]<-N20[1]
```

```
nTest[3]<-N20[2]
mHAG_Ref[3]<-BP20[1]
mHAG_Test[3]<-BP20[2]
diff[3]<-mHAG_Test[3]-mHAG_Ref[3]
# 25%
nRef[4]<-N25[1]
nTest[4]<-N25[2]
mHAG_Ref[4]<-BP25[1]
mHAG_Test[4]<-BP25[2]
diff[4]<-mHAG_Test[4]-mHAG_Ref[4]
# 30%
nRef[5]<-N30[1]
nTest[5]<-N30[2]
mHAG_Ref[5]<-BP30[1]
mHAG_Test[5]<-BP30[2]
diff[5]<-mHAG_Test[5]-mHAG_Ref[5]
# 50%
nRef[6]<-N50[1]
nTest[6]<-N50[2]
mHAG_Ref[6]<-BP50[1]
mHAG_Test[6]<-BP50[2]
diff[6]<-mHAG_Test[6]-mHAG_Ref[6]
# Collect info for each breakpoint and save
GLBEAST_BP<-data.frame(cbind(BPPercent,BPProps,      nRef,mHAG_
Ref,nTest,mHAG_Test,diff))
fnBP<-paste("./GLBEAST/Datasets/GLBEAST BP",glake,".Rds")
saveRDS(GLBEAST_BP,file=fnBP)
# Plot the difference between median HAG for Reference v Test against
breakpoint percentage
BPnlabels<-paste("nRef  =  ",as.character(nRef),"\n  nTest  =
",as.character(nTest))
d2<-cbind(GLBEAST_BP,BPnlabels)
ggplot(d2, aes(BPPercent,diff,label=BPnlabels)) +
theme(axis.text = element_text(color = "blue", size = 16)) +
theme(axis.title  =  element_text(color  =  "blue",  size  =
14,face="bold")) +
ggtitle(output_title) +
theme(title = element_text(color = "blue", size = 16,face="bold"))
+
xlim(10,50) +
xlab("Human Activity Gradient Percentile Breakpoint") +
ylab("Test - Ref Human Activity") +
geom_point() +
geom_line() +
geom_label(size=5)
```

**GLBEAST 2_4 Defining Reference w HAGs Lake Huron.R – For Lake Huron, calculates and plots difference in median Human Activity between Reference and Test sites for different breakpoints along human activity gradient.**

```
library(GGally)
library(tidyverse)
library(RColorBrewer)
```

```
# Load data for plotting
d<-readRDS("./GLBEAST/Datasets/GLBEAST Env w HAGs.Rds")
# Plot title
glake<-"Lake Huron"
output_title<-paste(glake)
output_subtitle<-("Defining Reference v Test")
# Pick the subset of observations
# Either use all (Line7) or use a subgroup
d<-subset(d,GreatLake==glake)
hist(d$HAG_PC1)
# Get the potential breakpoints along the HAG for defining Reference
v Test
BPPercent<-c(10,15,20,25,30,50)
BPProps<-BPPercent/100
GLBEAST_BP<-matrix(quantile(d$HAG_PC1, probs = BPProps),6,1)
# Create vectors to hold Status with different breakpoints and
Reference v Test Comparisons
Status10<-c(rep("R",nrow(d)))
Status15<-c(rep("R",nrow(d)))
Status20<-c(rep("R",nrow(d)))
Status25<-c(rep("R",nrow(d)))
Status30<-c(rep("R",nrow(d)))
Status50<-c(rep("R",nrow(d)))
# Create vectors to hold Reference v Test Comparisons for each break-
point
nRef<-c(rep(0,nrow(GLBEAST_BP)))
nTest<-c(rep(0,nrow(GLBEAST_BP)))
mHAG_Ref<-c(rep(0,nrow(GLBEAST_BP)))
mHAG_Test<-c(rep(0,nrow(GLBEAST_BP)))
diff<-mHAG_Test-mHAG_Ref
# Loop over sites to define Reference v Test for each possible break-
point
for(Site_row in 1:nrow(d)) {
Status10[Site_row]<-ifelse(d$HAG_PC1[Site_row]>GLBEAST_
BP[1],"T","R")
Status15[Site_row]<-ifelse(d$HAG_PC1[Site_row]>GLBEAST_
BP[2],"T","R")
Status20[Site_row]<-ifelse(d$HAG_PC1[Site_row]>GLBEAST_
BP[3],"T","R")
Status25[Site_row]<-ifelse(d$HAG_PC1[Site_row]>GLBEAST_
BP[4],"T","R")
Status30[Site_row]<-ifelse(d$HAG_PC1[Site_row]>GLBEAST_
BP[5],"T","R")
Status50[Site_row]<-ifelse(d$HAG_PC1[Site_row]>GLBEAST_
BP[6],"T","R")
}
# Append Status for each breakpoint to HAG data for each site and save
d<-cbind(d,Status10,Status15,Status20,Status25,Status30,Status50)
fnStatus<-paste("./GLBEAST/Datasets/GLBEAST w HAGs
Status",glake,".Rds")
saveRDS(d,file=fnStatus)
#Collect up medians and sample sizes from Reference and Test groups
for each breakpoint
# Medians
```

```
BP10<-tapply(d$HAG_PC1,d$Status10,median)
BP15<-tapply(d$HAG_PC1,d$Status15,median)
BP20<-tapply(d$HAG_PC1,d$Status20,median)
BP25<-tapply(d$HAG_PC1,d$Status25,median)
BP30<-tapply(d$HAG_PC1,d$Status30,median)
BP50<-tapply(d$HAG_PC1,d$Status50,median)
# Sample sizes
N10<-tapply(d$HAG_PC1,d$Status10,length)
N15<-tapply(d$HAG_PC1,d$Status15,length)
N20<-tapply(d$HAG_PC1,d$Status20,length)
N25<-tapply(d$HAG_PC1,d$Status25,length)
N30<-tapply(d$HAG_PC1,d$Status30,length)
N50<-tapply(d$HAG_PC1,d$Status50,length)
# Now build the table that will enable the best decision about break-
point
# 10%
nRef[1]<-N10[1]
nTest[1]<-N10[2]
mHAG_Ref[1]<-BP10[1]
mHAG_Test[1]<-BP10[2]
diff[1]<-mHAG_Test[1]-mHAG_Ref[1]
# 15%
nRef[2]<-N15[1]
nTest[2]<-N15[2]
mHAG_Ref[2]<-BP15[1]
mHAG_Test[2]<-BP15[2]
diff[2]<-mHAG_Test[2]-mHAG_Ref[2]
# 20%
nRef[3]<-N20[1]
nTest[3]<-N20[2]
mHAG_Ref[3]<-BP20[1]
mHAG_Test[3]<-BP20[2]
diff[3]<-mHAG_Test[3]-mHAG_Ref[3]
# 25%
nRef[4]<-N25[1]
nTest[4]<-N25[2]
mHAG_Ref[4]<-BP25[1]
mHAG_Test[4]<-BP25[2]
diff[4]<-mHAG_Test[4]-mHAG_Ref[4]
# 30%
nRef[5]<-N30[1]
nTest[5]<-N30[2]
mHAG_Ref[5]<-BP30[1]
mHAG_Test[5]<-BP30[2]
diff[5]<-mHAG_Test[5]-mHAG_Ref[5]
# 50%
nRef[6]<-N50[1]
nTest[6]<-N50[2]
mHAG_Ref[6]<-BP50[1]
mHAG_Test[6]<-BP50[2]
diff[6]<-mHAG_Test[6]-mHAG_Ref[6]
# Collect info for each breakpoint and save
GLBEAST_BP<-data.frame(cbind(BPPercent,BPProps,        nRef,mHAG_
Ref,nTest,mHAG_Test,diff))
```

```
fnBP<-paste("./GLBEAST/Datasets/GLBEAST BP",glake,".Rds")
saveRDS(GLBEAST_BP,file=fnBP)
# Plot the difference between median HAG for Reference v Test against
breakpoint percentage
BPnlabels<-paste("nRef = ",as.character(nRef),"\n nTest =
",as.character(nTest))
d2<-cbind(GLBEAST_BP,BPnlabels)
ggplot(d2, aes(BPPercent,diff,label=BPnlabels)) +
theme(axis.text = element_text(color = "blue", size = 16)) +
theme(axis.title = element_text(color = "blue", size =
14,face="bold")) +
ggtitle(output_title) +
theme(title = element_text(color = "blue", size = 16,face="bold"))
+
xlim(10,50) +
xlab("Human Activity Gradient Percentile Breakpoint") +
ylab("Test - Ref Human Activity") +
geom_point() +
geom_line() +
geom_label(size=5)
```

**GLBEAST 2_4 Defining Reference w HAGs Lake Erie.R – For Lake Erie, calculates and plots difference in median Human Activity between Reference and Test sites for different breakpoints along human activity gradient.**

```
library(GGally)
library(tidyverse)
library(RColorBrewer)
# Load data for plotting
d<-readRDS("./GLBEAST/Datasets/GLBEAST Env w HAGs.Rds")
# Plot title
glake<-"Lake Erie"
output_title<-paste(glake)
output_subtitle<-("Defining Reference v Test")
# Pick the subset of observations
# Either use all (Line7) or use a subgroup
d<-subset(d,GreatLake==glake)
hist(d$HAG_PC1)
# Get the potential breakpoints along the HAG for defining Reference
v Test
BPPercent<-c(10,15,20,25,30,50)
BPProps<-BPPercent/100
GLBEAST_BP<-matrix(quantile(d$HAG_PC1, probs = BPProps),6,1)
# Create vectors to hold Status with different breakpoints and
Reference v Test Comparisons
Status10<-c(rep("R",nrow(d)))
Status15<-c(rep("R",nrow(d)))
Status20<-c(rep("R",nrow(d)))
Status25<-c(rep("R",nrow(d)))
Status30<-c(rep("R",nrow(d)))
Status50<-c(rep("R",nrow(d)))
# Create vectors to hold Reference v Test Comparisons for each break-
point
```

```
nRef<-c(rep(0,nrow(GLBEAST_BP)))
nTest<-c(rep(0,nrow(GLBEAST_BP)))
mHAG_Ref<-c(rep(0,nrow(GLBEAST_BP)))
mHAG_Test<-c(rep(0,nrow(GLBEAST_BP)))
diff<-mHAG_Test-mHAG_Ref
# Loop over sites to define Reference v Test for each possible break-
point
for(Site_row in 1:nrow(d)) {
Status10[Site_row]<-ifelse(d$HAG_PC1[Site_row]>GLBEAST_
BP[1],"T","R")
Status15[Site_row]<-ifelse(d$HAG_PC1[Site_row]>GLBEAST_
BP[2],"T","R")
Status20[Site_row]<-ifelse(d$HAG_PC1[Site_row]>GLBEAST_
BP[3],"T","R")
Status25[Site_row]<-ifelse(d$HAG_PC1[Site_row]>GLBEAST_
BP[4],"T","R")
Status30[Site_row]<-ifelse(d$HAG_PC1[Site_row]>GLBEAST_
BP[5],"T","R")
Status50[Site_row]<-ifelse(d$HAG_PC1[Site_row]>GLBEAST_
BP[6],"T","R")
}
# Append Status for each breakpoint to HAG data for each site and save
d<-cbind(d,Status10,Status15,Status20,Status25,Status30,Status50)
fnStatus<-paste("./GLBEAST/Datasets/GLBEAST w HAGs
Status",glake,".Rds")
saveRDS(d,file=fnStatus)
#Collect up medians and sample sizes from Reference and Test groups
for each breakpoint
# Medians
BP10<-tapply(d$HAG_PC1,d$Status10,median)
BP15<-tapply(d$HAG_PC1,d$Status15,median)
BP20<-tapply(d$HAG_PC1,d$Status20,median)
BP25<-tapply(d$HAG_PC1,d$Status25,median)
BP30<-tapply(d$HAG_PC1,d$Status30,median)
BP50<-tapply(d$HAG_PC1,d$Status50,median)
# Sample sizes
N10<-tapply(d$HAG_PC1,d$Status10,length)
N15<-tapply(d$HAG_PC1,d$Status15,length)
N20<-tapply(d$HAG_PC1,d$Status20,length)
N25<-tapply(d$HAG_PC1,d$Status25,length)
N30<-tapply(d$HAG_PC1,d$Status30,length)
N50<-tapply(d$HAG_PC1,d$Status50,length)
# Now build the table that will enable the best decision about break-
point
# 10%
nRef[1]<-N10[1]
nTest[1]<-N10[2]
mHAG_Ref[1]<-BP10[1]
mHAG_Test[1]<-BP10[2]
diff[1]<-mHAG_Test[1]-mHAG_Ref[1]
# 15%
nRef[2]<-N15[1]
nTest[2]<-N15[2]
mHAG_Ref[2]<-BP15[1]
```

```
mHAG_Test[2]<-BP15[2]
diff[2]<-mHAG_Test[2]-mHAG_Ref[2]
# 20%
nRef[3]<-N20[1]
nTest[3]<-N20[2]
mHAG_Ref[3]<-BP20[1]
mHAG_Test[3]<-BP20[2]
diff[3]<-mHAG_Test[3]-mHAG_Ref[3]
# 25%
nRef[4]<-N25[1]
nTest[4]<-N25[2]
mHAG_Ref[4]<-BP25[1]
mHAG_Test[4]<-BP25[2]
diff[4]<-mHAG_Test[4]-mHAG_Ref[4]
# 30%
nRef[5]<-N30[1]
nTest[5]<-N30[2]
mHAG_Ref[5]<-BP30[1]
mHAG_Test[5]<-BP30[2]
diff[5]<-mHAG_Test[5]-mHAG_Ref[5]
# 50%
nRef[6]<-N50[1]
nTest[6]<-N50[2]
mHAG_Ref[6]<-BP50[1]
mHAG_Test[6]<-BP50[2]
diff[6]<-mHAG_Test[6]-mHAG_Ref[6]
# Collect info for each breakpoint and save
GLBEAST_BP<-data.frame(cbind(BPPercent,BPProps,      nRef,mHAG_
Ref,nTest,mHAG_Test,diff))
fnBP<-paste("./GLBEAST/Datasets/GLBEAST BP",glake,".Rds")
saveRDS(GLBEAST_BP,file=fnBP)
# Plot the difference between median HAG for Reference v Test against
breakpoint percentage
BPnlabels<-paste("nRef   =   ",as.character(nRef),"\n   nTest   =
",as.character(nTest))
d2<-cbind(GLBEAST_BP,BPnlabels)
ggplot(d2, aes(BPPercent,diff,label=BPnlabels)) +
theme(axis.text = element_text(color = "blue", size = 16)) +
theme(axis.title   =   element_text(color   =   "blue",   size   =
14,face="bold")) +
ggtitle(output_title) +
theme(title = element_text(color = "blue", size = 16,face="bold"))
+
xlim(10,50) +
xlab("Human Activity Gradient Percentile Breakpoint") +
ylab("Test - Ref Human Activity") +
geom_point() +
geom_line() +
geom_label(size=5)
```

**GLBEAST 2_4 Defining Reference w HAGs Lake Ontario.R – For Lake Ontario, calculates and plots difference in median Human Activity between Reference and Test sites for different breakpoints along human activity gradient.**

```
library(GGally)
library(tidyverse)
library(RColorBrewer)
# Load data for plotting
d<-readRDS("./GLBEAST/Datasets/GLBEAST Env w HAGs.Rds")
# Plot title
glake<-"Lake Ontario"
output_title<-paste(glake)
output_subtitle<-("Defining Reference v Test")
# Pick the subset of observations
# Either use all (Line7) or use a subgroup
d<-subset(d,GreatLake==glake)
hist(d$HAG_PC1)
# Get the potential breakpoints along the HAG for defining Reference
v Test
BPPercent<-c(10,15,20,25,30,50)
BPProps<-BPPercent/100
GLBEAST_BP<-matrix(quantile(d$HAG_PC1, probs = BPProps),6,1)
# Create vectors to hold Status with different breakpoints and
Reference v Test Comparisons
Status10<-c(rep("R",nrow(d)))
Status15<-c(rep("R",nrow(d)))
Status20<-c(rep("R",nrow(d)))
Status25<-c(rep("R",nrow(d)))
Status30<-c(rep("R",nrow(d)))
Status50<-c(rep("R",nrow(d)))
# Create vectors to hold Reference v Test Comparisons for each break-
point
nRef<-c(rep(0,nrow(GLBEAST_BP)))
nTest<-c(rep(0,nrow(GLBEAST_BP)))
mHAG_Ref<-c(rep(0,nrow(GLBEAST_BP)))
mHAG_Test<-c(rep(0,nrow(GLBEAST_BP)))
diff<-mHAG_Test-mHAG_Ref
# Loop over sites to define Reference v Test for each possible break-
point
for(Site_row in 1:nrow(d)) {
Status10[Site_row]<-ifelse(d$HAG_PC1[Site_row]>GLBEAST_
BP[1],"T","R")
Status15[Site_row]<-ifelse(d$HAG_PC1[Site_row]>GLBEAST_
BP[2],"T","R")
Status20[Site_row]<-ifelse(d$HAG_PC1[Site_row]>GLBEAST_
BP[3],"T","R")
Status25[Site_row]<-ifelse(d$HAG_PC1[Site_row]>GLBEAST_
BP[4],"T","R")
Status30[Site_row]<-ifelse(d$HAG_PC1[Site_row]>GLBEAST_
BP[5],"T","R")
Status50[Site_row]<-ifelse(d$HAG_PC1[Site_row]>GLBEAST_
BP[6],"T","R")
}
# Append Status for each breakpoint to HAG data for each site and save
d<-cbind(d,Status10,Status15,Status20,Status25,Status30,Status50)
fnStatus<-paste("./GLBEAST/Datasets/GLBEAST w HAGs
Status",glake,".Rds")
saveRDS(d,file=fnStatus)
```

```
#Collect up medians and sample sizes from Reference and Test groups
for each breakpoint
# Medians
BP10<-tapply(d$HAG_PC1,d$Status10,median)
BP15<-tapply(d$HAG_PC1,d$Status15,median)
BP20<-tapply(d$HAG_PC1,d$Status20,median)
BP25<-tapply(d$HAG_PC1,d$Status25,median)
BP30<-tapply(d$HAG_PC1,d$Status30,median)
BP50<-tapply(d$HAG_PC1,d$Status50,median)
# Sample sizes
N10<-tapply(d$HAG_PC1,d$Status10,length)
N15<-tapply(d$HAG_PC1,d$Status15,length)
N20<-tapply(d$HAG_PC1,d$Status20,length)
N25<-tapply(d$HAG_PC1,d$Status25,length)
N30<-tapply(d$HAG_PC1,d$Status30,length)
N50<-tapply(d$HAG_PC1,d$Status50,length)
# Now build the table that will enable the best decision about break-
point
# 10%
nRef[1]<-N10[1]
nTest[1]<-N10[2]
mHAG_Ref[1]<-BP10[1]
mHAG_Test[1]<-BP10[2]
diff[1]<-mHAG_Test[1]-mHAG_Ref[1]
# 15%
nRef[2]<-N15[1]
nTest[2]<-N15[2]
mHAG_Ref[2]<-BP15[1]
mHAG_Test[2]<-BP15[2]
diff[2]<-mHAG_Test[2]-mHAG_Ref[2]
# 20%
nRef[3]<-N20[1]
nTest[3]<-N20[2]
mHAG_Ref[3]<-BP20[1]
mHAG_Test[3]<-BP20[2]
diff[3]<-mHAG_Test[3]-mHAG_Ref[3]
# 25%
nRef[4]<-N25[1]
nTest[4]<-N25[2]
mHAG_Ref[4]<-BP25[1]
mHAG_Test[4]<-BP25[2]
diff[4]<-mHAG_Test[4]-mHAG_Ref[4]
# 30%
nRef[5]<-N30[1]
nTest[5]<-N30[2]
mHAG_Ref[5]<-BP30[1]
mHAG_Test[5]<-BP30[2]
diff[5]<-mHAG_Test[5]-mHAG_Ref[5]
# 50%
nRef[6]<-N50[1]
nTest[6]<-N50[2]
mHAG_Ref[6]<-BP50[1]
mHAG_Test[6]<-BP50[2]
diff[6]<-mHAG_Test[6]-mHAG_Ref[6]
```

```
# Collect info for each breakpoint and save
GLBEAST_BP<-data.frame(cbind(BPPercent,BPProps,      nRef,mHAG_
Ref,nTest,mHAG_Test,diff))
fnBP<-paste("./GLBEAST/Datasets/GLBEAST BP",glake,".Rds")
saveRDS(GLBEAST_BP,file=fnBP)
# Plot the difference between median HAG for Reference v Test against
breakpoint percentage
BPnlabels<-paste("nRef   =   ",as.character(nRef),"\n  nTest   =
",as.character(nTest))
d2<-cbind(GLBEAST_BP,BPnlabels)
ggplot(d2, aes(BPPercent,diff,label=BPnlabels)) +
theme(axis.text = element_text(color = "blue", size = 16)) +
theme(axis.title   =   element_text(color   =   "blue",   size   =
14,face="bold")) +
ggtitle(output_title) +
theme(title = element_text(color = "blue", size = 16,face="bold"))
+
xlim(10,50) +
xlab("Human Activity Gradient Percentile Breakpoint") +
ylab("Test - Ref Human Activity") +
geom_point() +
geom_line() +
geom_label(size=5)
```

## References


Bailey, R.C., Green, R.H.: Within-basin variation in the shell morphology and growth-rate of a fresh-water mussel. Can. J. Zool. **66**, 1704–1708 (1988)

Bailey, R.C., Norris, R.H., Reynoldson, T.B.: Bioassessment of Freshwater Ecosystems-Using the Reference Condition Approach. Kluwer Academic Publishers, Boston (2004)

Bailey, R.C., Linke, S., Yates, A.G.: Bioassessment of freshwater ecosystems using the reference condition approach: comparing established and new methods with common data sets. Freshw. Sci. **33**, 1204–1211 (2014)

Davies, S.P., Jackson, S.K.: The biological condition gradient: a descriptive model for interpreting change in aquatic ecosystems. Ecol. Appl. **16**, 1251–1266 (2006)

Hynes, H.B.N.: The stream and its valley. Verh. Internat. Verein. Limnol. **19**, 1–15 (1975)

MacArthur, R.H., Wilson, E.O.: Equilibrium-theory of insular zoogeography. Evolution **17**, 373–0 (1963)

Mason, L.A., Riseng, C.M., Layman, A.J., Jensen, R.: Effective fetch and relative exposure index maps for the Laurentian Great Lakes. Sci Data **5**, 180295 (2018)

McElreath, R.: Statistical Rethinking: A Bayesian Course with Examples in R and STAN. 2nd edition. Chapman & Hall/CRC (2022)

Stoddard, J.L., Larsen, D.P., Hawkins, C.P., Johnson, R.K., Norris, R.H.: Setting expectations for the ecological condition of streams: the concept of reference condition. Ecol. Appl. **16**, 1267–1276 (2006)

Yates, A.G., Culp, J.M., Bailey, R.C., Chambers, P.A.: Strengths and weaknesses of data sources for describing exposure of aquatic ecosystems to human activity. Am. Fish. Soc. Symp. **90**, 11–33 (2019)

Yates, A.G., Bailey, R.C.: Selecting objectively defined reference sites for stream bioassessment programs. Environ. Monit. Assess. **170** (2010b)

# Chapter 3
# Modeling Variation in the Biota Among Reference Sites

## Why Isn't Everything Everywhere?

There are up to 25,000 blue whales living on the planet as I write this in 2025. The Earth's surface area is 510 million square kilometers, which is about 5 whales in every block of 10,000 $km^2$, but it doesn't take a marine mammologist to know that they aren't distributed uniformly or randomly among those blocks. My marine ecology colleagues (or my 10 year old grandkid) might gently remind me that blue whales just live in the oceans, which are about 71% (or 361 million square kilometers) of the Earth's surface, so I can refine my estimate to about 7 whales in each block of 10,000 $km^2$ *of ocean.* I can increase my precision even more if I do a bit of reading and recognize that blue whales are found in all except the Arctic Ocean, and they tend to migrate in response to food availability, often heading toward the equator as winter approaches in the southern or northern hemispheres (Anonymous 2024). If we built a model to explain the distribution and abundance of blue whales around the world, some of its parameters would be so obvious that they needn't be specified (e.g., the whales live in the ocean). Some parameters would be useful and generalizable (e.g., time of year, distance from the coast) and some would be perhaps too fine-grained to be useful in answering the question (e.g., distance from submarines or supertankers affect whale behavior). So, like every organism, the blue whale has a non-random distribution, and this obvious fact… obvious to even the most casual student of ecology… holds at temporal scales from milliseconds to millennia and spatial scales from millimeters to globally.

In my more than 35 years as a biology professor, I have periodically taught ecology to students in their first or second year of university. Because these students are often more interested in a future as a doctor or a dentist than as an environmental scientist, I've tried to express ecological ideas as clearly and concisely as possible. This has brought me collateral benefits… teaching often clarifies ideas I wrestle with in research, since there is nothing like figuring out how to explain something to make it absolutely clear in your own head. On the first day of class, I share my definition of

 
R. C. Bailey, *Rethinking Bioassessment of Freshwater Ecosystem Health*,
https://doi.org/10.1007/978-3-032-12427-2_3

ecology by asking the students to think about their favorite ecosystem. Some of them think about a patch of forest on the edge of a field, or an oak tree in a park, but for me it's always a stream near my cottage where I walk the dog. I ask the students to think about what does and doesn't live in their favorite ecosystem and tell them that the central question asked by ecologists, in a huge diversity of ways and places, is "Why isn't everything everywhere?" In other words, every species from every kingdom has a certain global abundance, but we certainly don't find them randomly distributed on the globe. There isn't much chance that I'll find a blue whale in my local stream!

Once I have the students thinking about their favorite ecosystem, and the fact that "everything isn't everywhere", I ask them to really think about *why* there is a particular set of species in a certain ecosystem at a certain time. It really comes down to just three factors:

- **Getting there**—The species had to at some point travel or be transported to the ecosystem.
- **Being there**—The physical and chemical conditions (e.g., temperature and oxygen concentration) have to be within the range that the species can tolerate.
- **Staying there**—The species must be able to survive in the face of the abundance of competitors, predators, parasites, resources, and commensals that are also present in the ecosystem.

We can integrate basic ecology with environmental assessment by applying the logic of bioassessment that I described in Chap. 1. Each of the factors that determine whether or not a species is present in an ecosystem can be affected by a variety of both the natural and human activity environments. Our job becomes describing an ecosystem's natural and human activity environments, as well as its biota, precisely enough to understand:

- **Variation in the biota among Reference sites** How does variation in the natural environment ***cause*** variation in the ecosystem's biota (this chapter)?
- **Bioassessment** Is the biota we find at a Test site what we would expect if the ecosystem was in Reference Condition (Chap. 4)?
- **Diagnosis** How does the biota at a site vary in relation to the human activity environment of the ecosystem (Chap. 4)?

In Chap. 2, I brought these three points together in a ***conceptual model*** for a research or bioassessment program. Conceptual models just include the factors that I ***think*** are affecting the ecosystems of interest. It doesn't include the magnitude or even sign of the effect, and it doesn't include interactions among these factors that may be present. I'm not advocating the kind of game some play in setting up statistical hypothesis tests before looking at the data… in fact, I'm not particularly interested in statistical hypothesis tests anymore! I'd rather stick to building models which include variables where I have some notion of a causal link between predictors and responses, with due consideration of potentially confounding variables. I got inspired to think deeply about this from a great book by Richard McElreath (McElreath 2020), and I've tried to illustrate the value of conceptual models with my two example datasets (SimStreams in Fig. 2.1, GLBEAST in Fig. 2.3). If you start by thinking through the conceptual model, it forces you to think clearly about what really drives variation in the biota from one site to the next.

## Describing the Biota

After sorting out the conceptual model, I usually take a look at biological data from just Reference sites to start to get to know the dataset, and then I start building and comparing candidate models of the Reference Condition. But first I have to decide what descriptors of the biota to model. As I discussed in Chap. 1, heated debate about this has consumed bioassessment scientists for decades, although we seem to be at a relatively peaceful phase these days. The main issue now is how much of the diversity in the biota we miss with morpho-taxonomy and how much value and credibility quantitative estimates of each taxon's numbers or biomass really has.

Much of the angst in the past, I think, was the notion that there was one particular descriptor, whether a "multi-metric" that brought several tolerance and diversity indices together or a multivariate characterization of a community in an ordination, that was "best". I think part of this is entangled with the NHST mindset of a lot of bioassessment work with decisions about a Test site ***passing*** or ***failing*** and **Type I** (failing a site in Reference Condition) and **Type II** (passing a site not in Reference Condition) **errors**. Well I've been there and done that, but perhaps because of my stage of life and career I'm less attracted to a decision dichotomy and more amenable to a narrative report on a Test site. This approach to bioassessment is a little less clear cut but I think gets closer to what we can and can't say based on our data and our models.

To illustrate my approach with Bayesian fitting of both Reference Condition (this chapter) and diagnosis (Chap. 4) models, I use the very simple **taxon richness** (species richness for the SimStreams dataset) and the multivariate characterization of a community from presence/absence data with **ordination scores**. Note that I am avoiding a lot of angst about whether to use "quantitative" (abundance or biomass per unit area or volume, or relative abundance or biomass of the taxa) or "qualitative" (was it there or not) data. I'm not claiming that it's not an important discussion…I just feel like it gets in the way of clarity in the bioassessment approach. Taxon richness is just how many taxa (at whatever taxonomic level is used) were detected in the sampling. Ordination scores are often more obscure, especially to folks actually implementing or trying to understand bioassessment programs. I just think of them as a description of the community composition. Two sites that have similar ordination scores will have similar community composition and vice versa!

## Modeling the Reference Condition

In this chapter, I will build a model of the effects of the natural environment on the biota just using the Reference sites we defined in Chap. 2. In Chap. 4, I will assess the likelihood of Test sites having the biota we expect from our Reference Condition model. I will then use both Reference and Test sites to see if there is a relationship between the biota and human activity so that I can build at least a circumstantial case that human activity *causes* Test sites to be out of Reference Condition.

## *SimStreams*

So now, finally, we will model the Reference Condition in SimStreams. We'll use this model later to predict the biota we expect to see at a Test site with a given natural environment if it is in Reference Condition. Using the conceptual model for SimStreams (Fig. 2.1), we'll determine the very best model that predicts the biota from the natural environment of Reference sites (Ecoregion, Catchment Area). Our approach will be quite different from the usual methods. No classification of Reference sites based on the biota followed by discriminant analysis of the resulting biota-defined groups using natural environment descriptors. No building simple or complex linear models where we use some stepwise model-building algorithm and then test hypotheses about coefficients. Instead, we will fit a variety of models using a Bayesian approach, and then choose the best model that optimally combines fit and efficiency. Bayesian model-fitting sounds intimidating to those who haven't tried it, but it's really just finding the most likely alternatives when there are a sometimes very large number of options. It is best illustrated with examples, so let's fit and then compare a number of Reference Condition models.

### Species Richness

Each of my simulated Reference ecosystems has a particular value of species richness. The first thing I always do before I get into the modeling is take a look at the data. When I do simple boxplots of species richness in Reference sites in each of the two Ecoregions (Fig. 3.1), the difference is pretty striking and makes sense. There are more species in the Valley Ecoregion…there are a greater variety of niches and more nutrients in the Valley streams than in the Highland streams.

The simplest, "null" model (we'll call it **RC0_Null**) is that species richness of the Reference sites is from a normal distribution with mean $\mu$ and standard deviation $\sigma$. A normal distribution seems reasonable…many factors combined together that determine the number of species at a given site tend to generate a normal distribution among sites. But this probably seems weird to my fellow bioassessment scientists. It's kind of like modeling blue whale distribution as a normal distribution among

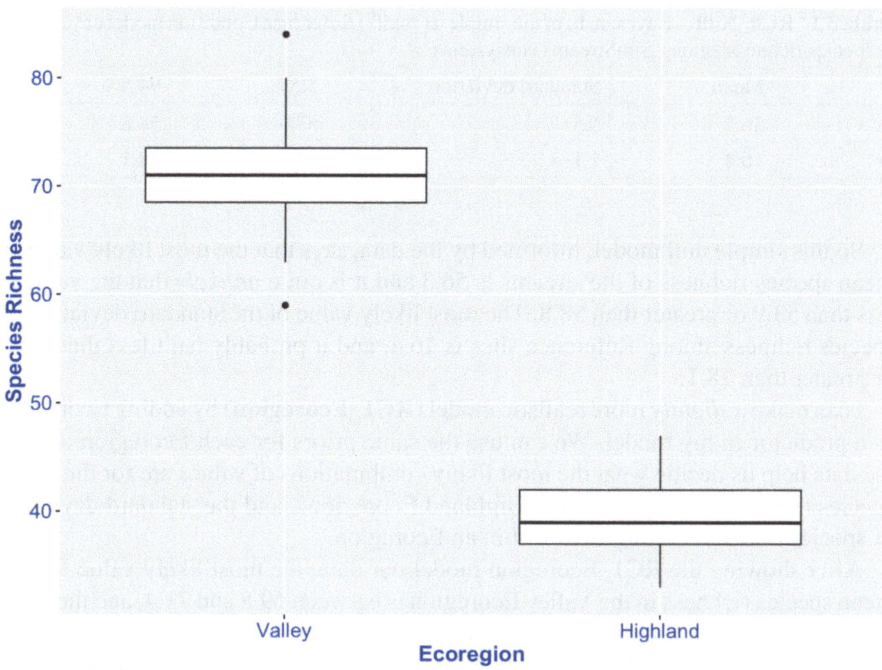


**Fig. 3.1** Species richness of Reference sites in Valley (n = 63) and Highland (n = 50) Ecoregions

the blocks of 10,000 km$^2$ in the oceans without any other predictors. What it's saying in statistical terms is that the species richness I see at each site is from a normal distribution with mean $\mu$ and a standard deviation among streams of $\sigma$. Bayesian fitting is going to try a whole bunch of possible values of $\mu$ and $\sigma$ and see which combination is more or less likely given the data we actually got. And for bioassessment, if this ends up being the best model of the Reference site biota (unlikely), we can find the most likely values of $\mu$ and $\sigma$ and then identify Test sites that have unusually high or low values of species richness relative to the expected value.

This gets to why I like Bayesian fitting versus the null hypothesis and significance testing (NHST), frequentist approach, which just tells me the chance of getting the data I got if the null hypothesis is true. With Bayesian fitting, rather than a "null, yes or no?" process I get the likelihood of an infinite number of possibilities. The range of combinations of $\mu$ and $\sigma$ values we try out are called "priors", and in fitting this simple null model, I used priors for $\mu$ with normally distributed means of 50 species and a standard error of the mean of ten species, and among-site standard deviation values uniformly distributed from 0 to 25. The fitting process samples values for $\mu$ and $\sigma$ from this priors distribution and uses the data to figure out the likelihood of this set of values. Lots and lots of sampling of the priors is done, resulting in the ***most likely*** values of $\mu$ and $\sigma$ in Table 3.1.

**Table 3.1** **RC0_Null**—Bayesian fit of the simplest ("null") Reference Condition model of variation in species richness among SimStreams ecosystems

| | Mean | Standard deviation | 5.5% | 94.5% |
|---|---|---|---|---|
| $\mu$ | 56.3 | 1.5 | 53.9 | 58.8 |
| $\sigma$ | 16.4 | 1.1 | 14.6 | 18.1 |

So this simple null model, informed by the data, says that the most likely value of mean species richness of the streams is 56.3 and it is quite *unlikely* that the value is less than 53.9 or greater than 58.8. The most likely value of the standard deviation of species richness among Reference sites is 16.4, and it probably isn't less than 14.6 or greater than 18.1.

I can make a *slightly* more realistic model (**RC1_Ecoregion**) by adding Ecoregion as a predictor in my model. We can use the same priors for each Ecoregion and let the data help us decide what the most likely combinations of values are for the mean species richness in the Valley and Highland Ecoregions, and the standard deviation of species richness among sites within an Ecoregion.

After showing the RC1_Ecoregion model our data, the most likely value for the mean species richness in the Valley Ecoregion is between 69.8 and 71.4, and the most likely value for the Highland Ecoregion is between 37.8 and 39.5 (Table 3.2). Keep in mind that we are fitting a model here rather than determining confidence limits for an estimate of the "population means" or doing a t-test of differences between the Ecoregions. Note how much our mean likely value for $\sigma$ has shrunk from RC0_Null's $\sigma = 16.4$ to RC1_Ecoregion's $\sigma = 3.8$. Our refined model has definitely added some predictive power! If we were to do a bioassessment using this model, we'd now be expecting species richness of about 70 in a Valley Test stream and 40 in a Highland Test stream if they're in Reference Condition, and our standard deviation of 3 or 4 would tell us how unusually high or low the actual species richness we saw was.

My conceptual model (Fig. 2.1) also includes Catchment Area as a potentially important causal factor that helps describe the natural environment of the stream ecosystem. I can add that to the model with the same effect of Catchment Area for each Ecoregion (**RC2_Ecoregion_CA**). We can use the same priors as RC1_Ecoregion for each Ecoregion by trying out slope values for the relationship between Catchment Area and species richness with a mean and standard deviation of 1, and then once again let the data help us decide what the most (and least) likely

**Table 3.2** **RC1_Ecoregion**—Bayesian fit of the Ecoregion only Reference Condition model of variation in species richness among SimStreams ecosystems

| | Mean | Standard deviation | 5.5% | 94.5% |
|---|---|---|---|---|
| $\alpha_{valley}$ | 70.6 | 0.5 | 69.8 | 71.4 |
| $\alpha_{highland}$ | 38.6 | 0.5 | 37.8 | 39.5 |
| $\sigma$ | 3.8 | 0.2 | 3.4 | 4.2 |

combinations of values are the mean species richness in each Ecoregion, the slope relating Catchment Area to species richness, and the standard deviation among sites in a given Ecoregion at a given Catchment Area.

This time our likely value for $\sigma$ has shrunk more modestly from 3.8 in RC1_Ecoregion (Table 3.2) to σ = 3.1 in RC2_Ecoregion_CA (Table 3.3). The likely slope value is small (0.017–0.027), but keep in mind it is an estimate of how much richness increases with every extra hectare in the Catchment Area, so a 1000 more hectares in the Catchment Area means 22 more species (according to the model!).

Sometimes it is more appropriate to model an arithmetic change in the biota with a *percentage* change in the natural environment, particular if the natural environment descriptor is a measure of the size of the ecosystem. To do this, we can fit the same model as RC2_Ecoregion_CA but using log(CA) as a predictor (**RC3_Ecoregion_logCA**).

This time our likely value for $\sigma$ has increased to $\sigma = 3.4$ (Table 3.4) relative to RC2_Ecoregion_CA's $\sigma = 3.1$ (Table 3.3), indicative of a poorer fit to the data. The likely $\beta$ value is much bigger but that is just reflective of the log scale of Catchment Area.

The final model we will fit (**RC4_Ecoregion_CA_Interaction**) tries out a different linear effect of Catchment Area in each Ecoregion…in other words, an ***interaction*** between Ecoregion and Catchment Area in our model. This seems reasonable, given our conceptual model. Bigger catchments in the Valley Ecoregion may gain species more than they do in the Highlands … only so many species are available to colonize and can maintain a population in the Highlands.

In the fit of RC4_Ecoregion_CA_Interaction (Table 3.5), there is lots of overlap between likely values of $\beta_{valley}$ and $\beta_{highland}$, indicating little evidence of different relationships between Catchment Area and species richness in the two Ecoregions.

**Table 3.3 RC2_Ecoregion_CA**—Bayesian fit of the Ecoregion and Catchment Area Reference Condition model of variation in species richness among SimStreams ecosystems

| | Mean | Standard deviation | 5.5% | 94.5% |
|---|---|---|---|---|
| $\alpha_{valley}$ | 67.7 | 0.6 | 66.8 | 68.6 |
| $\alpha_{highland}$ | 42.4 | 0.7 | 41.3 | 43.4 |
| $\beta\_CA$ | 0.022 | 0.003 | 0.017 | 0.027 |
| $\sigma$ | 3.1 | 0.2 | 2.8 | 3.4 |

**Table 3.4 RC3_Ecoregion_logCA**—Bayesian fit of the Ecoregion and log(Catchment Area) Reference Condition model of variation in species richness among SimStreams ecosystems

| | Mean | Standard deviation | 5.5% | 94.5% |
|---|---|---|---|---|
| $\alpha_{valley}$ | 69.1 | 0.5 | 68.2 | 70.0 |
| $\alpha_{highland}$ | 40.5 | 0.6 | 39.5 | 41.5 |
| $\beta\_logCA$ | 3.09 | 0.65 | 2.06 | 4.1 |
| $\sigma$ | 3.4 | 0.2 | 3.0 | 3.8 |

The value of $\sigma$ (3.1) is also very close to that of RC2_Ecoregion_CA (Table 3.3), indicating no great increase in fit of the data to the model with this extra coefficient.

We have several competing models that all explain variation in biota among Reference sites in the SimStreams dataset. Choosing the best model, whether or not you use Bayesian fitting, is always tricky, always a trade-off between going with the best fit (e.g., with the lowest AIC, or the highest $R^2$, or all "significant" predictors) and a robust, generalizable model that helps us better understand the causal mechanisms in the ecosystems we are assessing. McElreath (2020, Chap. 7) has great guidance when it comes to choosing among competing models. He justifies using the Widely Applicable Information Criterion (WAIC) as a measure that effectively guards against overfitting data while picking a model that makes use of good predictors from your conceptual model. This and other diagnostics show that RC2_Ecoregion_CA, with Ecoregion and one coefficient for the linear effect of Catchment Area in both Ecoregions, is the best model (Table 3.6).

Before I do the bioassessment (Chap. 4), I find it really useful to show this model, along with the actual Reference and Test site data, as kind of a reality check (Fig. 3.2). Notice that many Test sites fall below the expected line in the Valley Ecoregion (less species richness than expected) and above the line in the Highland Ecoregion (higher species richness than expected).

**Table 3.5** **RC4_Ecoregion_CA_Interaction**—Bayesian fit of the Ecoregion, Catchment Area, and their interaction Reference Condition model of variation in species richness among SimStreams ecosystems

| | Mean | Standard deviation | 5.5% | 94.5% |
|---|---|---|---|---|
| $\alpha_{valley}$ | 67.2 | 0.6 | 66.3 | 68.3 |
| $\beta\_CA_{valley}$ | 0.025 | 0.004 | 0.019 | 0.031 |
| $\alpha_{highland}$ | 41.3 | 1.0 | 39.7 | 42.9 |
| $\beta\_CA_{highland}$ | 0.016 | 0.005 | 0.007 | 0.025 |
| $\sigma$ | 3.1 | 0.2 | 2.8 | 3.4 |

**Table 3.6** Comparing Reference Condition models of variation in species richness among SimStreams ecosystems

| | WAIC | SE | dWAIC | dSE | pWAIC | Weight |
|---|---|---|---|---|---|---|
| RC2_Ecoregion_CA | 585.7 | 15.37 | 0.0 | NA | 4.2 | 0.56 |
| RC4_Ecoregion_CA_Interaction | 586.2 | 16.09 | 0.4 | 2.89 | 5.3 | 0.44 |
| RC3_Ecoregion_logCA | 604.0 | 15.42 | 18.2 | 8.52 | 3.7 | 0.00 |
| RC1_Ecoregion | 625.4 | 15.34 | 39.7 | 12.41 | 2.9 | 0.00 |
| RC0_Null | 955.4 | 5.14 | 369.7 | 15.95 | 1.2 | 0.00 |

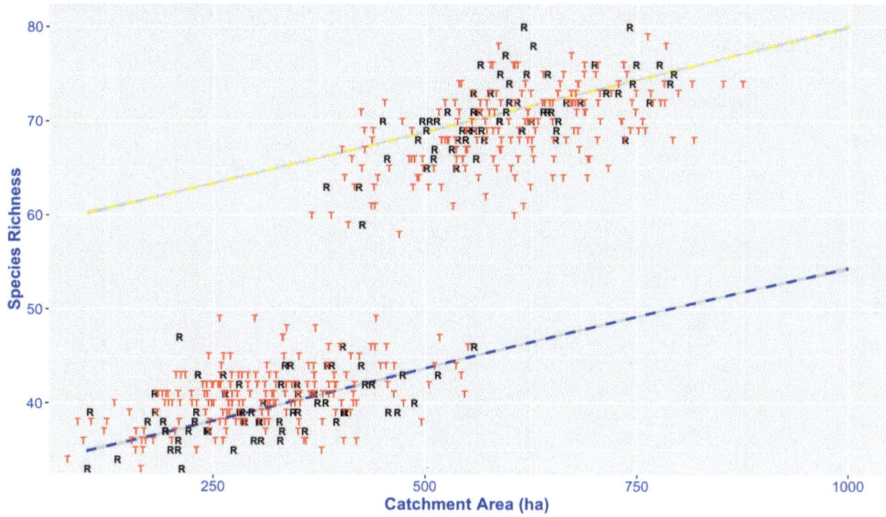


**Fig. 3.2** SimStreams species richness: RC2_Ecoregion_CA fit of Reference sites in Valley (n = 63; yellow line) and Highland (n = 50; blue line) Ecoregions. Actual (simulated) data from Reference (black "R") and Test (red "T") sites are also plotted

## Community Composition

When it comes to using ordination to show variation in community composition, I find that people tend to accept it without question as the best biota descriptor in bioassessment or cynically reject it as smoke and mirrors. Keep in mind that ordination is just a sketch (with some loss of information) of how much two communities differ…the distance between two sites on an ordination plot is a measure of how different their two communities are. I find it really useful plot the ordination scores but also, when I'm dealing with real data like GLBEAST, report actual taxon occurrences in addition to ordination scores. In spite of considerable variation in community composition within each Ecoregion, there does seem to be differences in composition between the Highland and Valley streams (Fig. 3.3).

I ran the same set of models as I did for species richness, and because the ordination scores were standardized to an overall mean of zero and standard deviation of one, I fit my null model with priors for $\mu$ with a normally distributed means of 1 and a standard error of the mean of 3 for both NMDS1 and NMDS2, and among site standard deviation values uniformly distributed from 0 to 3.

As with species richness, the best model for both NMDS1 and NMDS2 was RC2_Ecoregion_CA, which includes Ecoregion and Catchment Area as independent predictors of the ordination scores (Table 3.7).

The best models for NMDS1 and NMDS2, along with actual values for Reference and Test sites, are shown in Figs. 3.4 (NMDS1) and Fig. 3.5 (NMDS2). They show a composition change with Catchment Area, and we'll further examine the difference between Reference and Test sites in Chap. 4.

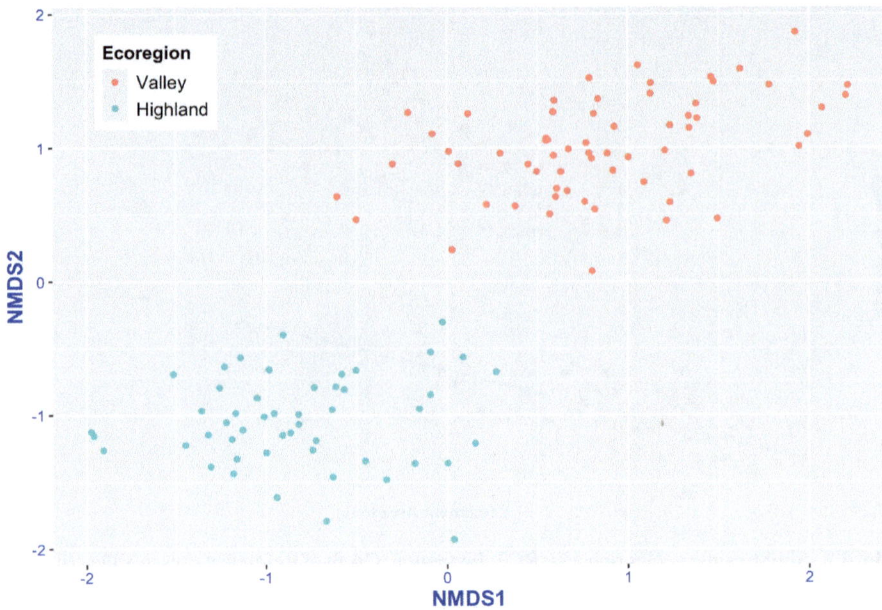


**Fig. 3.3** Community composition of Reference sites in Valley (n = 63) and Highland (n = 50) Ecoregions

**Table 3.7** Bayesian fit of the Ecoregion and Catchment Area Reference Condition model of variation in ordination scores among SimStreams ecosystems

| NMDS1 | Mean | Standard deviation | 5.5% | 94.5% |
|---|---|---|---|---|
| $\alpha_{valley}$ | 0.642 | 0.104 | 0.475 | 0.808 |
| $\alpha_{highland}$ | −0.793 | 0.066 | −0.899 | −0.687 |
| $\beta$ | 0.0017 | 0.0006 | 0.0008 | 0.0026 |
| $\sigma$ | 0.572 | 0.038 | 0.511 | 0.633 |
| NMDS2 | | | | |
| $\alpha_{valley}$ | 0.784 | 0.058 | 0.692 | 0.876 |
| $\alpha_{highland}$ | −0.760 | 0.069 | −0.870 | −0.650 |
| $\beta$ | 0.0017 | 0.0003 | 0.0012 | 0.0022 |
| $\sigma$ | 0.316 | 0.021 | 0.0283 | 0.350 |

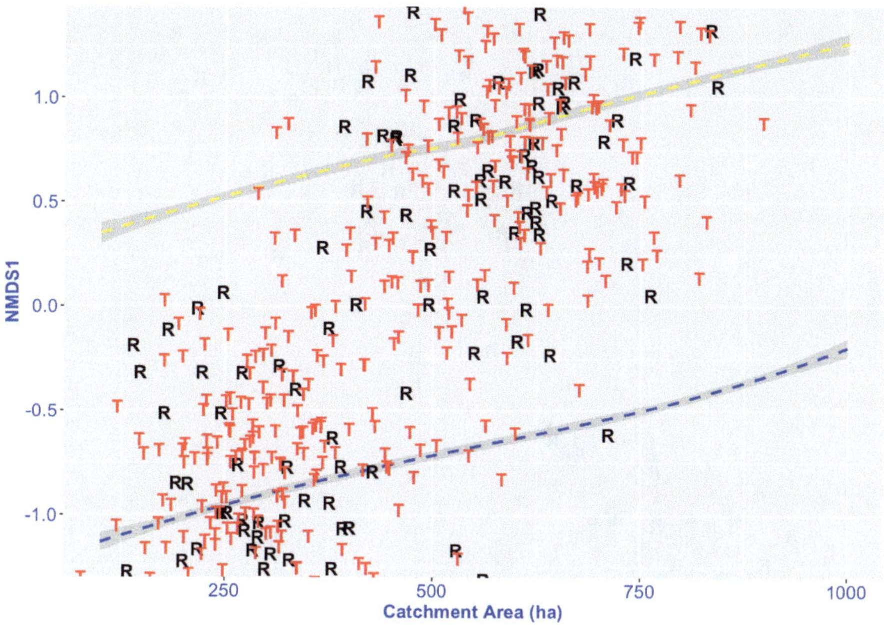


**Fig. 3.4** RC2_Ecoregion_CA fit of zNMDS1 for Reference sites in Valley (n = 63; yellow line) and Highland (n = 50; blue line) Ecoregions. Actual (simulated) data from Reference (black "R") and Test (red "T") sites also plotted

## GLBEAST

### *Taxon Richness*

There is some obvious variation among the Great Lakes Reference sites in taxon richness at a site, with Lake Huron the lowest site richness with many outliers and Lake Erie highest but quite variable (Fig. 3.6). The richness at a site (and other indices) should always be compared to the total richness across a group of sites to contrast $\propto$ (within-site) diversity with $\beta$ (site-to-site within a lake) diversity. I use a simple R script to calculate taxon richness and other diversity measures at the lake scale. The simplest way to calculate $\beta$ diversity is to divide the total number of taxa by the median number observed at a site: the larger this is, the more community composition varies from site to site (i.e., higher $\beta$ diversity). A simple example might help make this clear…if there are 10 taxa found at each site in a lake, and overall there are only 10 taxa found anywhere in the lake, then we're seeing the same community at every single site ($\beta$ diversity is 10/10 = 1. Alternatively, if there is a median of only two taxa found at a site, and there are 10 taxa found in the lake as a whole, the community varies a lot from site to site ($\beta$ diversity is 10/2 = 5). In the GLBEAST dataset, Lake Superior Reference sites had a total of 65 taxa and a median of 6 taxa

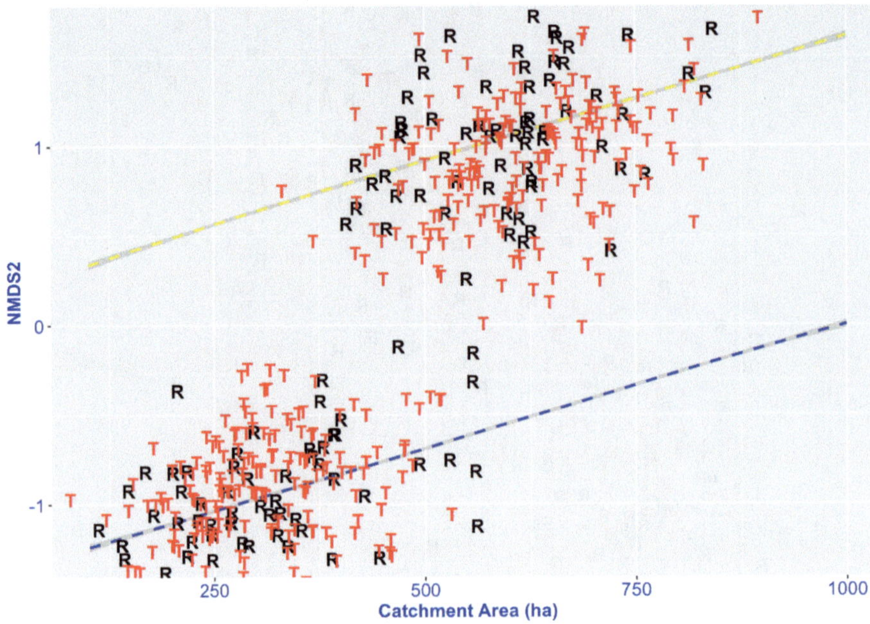


**Fig. 3.5** RC2_Ecoregion_CA fit of zNMDS2 for Reference sites in Valley (n = 63; yellow line) and Highland (n = 50; blue line) Ecoregions. Actual (simulated) data from Reference (black "R") and Test (red "T") sites also plotted

per site so $\beta$ diversity is 65/6 = 10.8, while Lake Huron had similar $\beta$ diversity of 70/7 = 10.0, Lake Erie had much lower $\beta$ diversity of 52/16.5 = 3.2, and Lake Ontario had $\beta$ diversity of 53/8 = 6.6.

As with SimStreams, the simplest model for taxon richness with the GLBEAST dataset (**RC0_Null**) is that taxon richness of each Reference site is from a normal distribution with mean $\mu$ and standard deviation $\sigma$. I fit RC0_Null for GLBEAST with priors from a normal distribution with mean 20 and standard error of the mean 10, and standard deviation values from 0 to 10. When this model is "shown" the data, the most likely number of taxa at each site is between 8 and 9, with a standard deviation among sites of between 5 and 6 (Table 3.8).

A slightly more realistic model (**RC1_GreatLake**) includes each Reference site's Great Lake as a predictor. There is good evidence (Table 3.9) that adding the Great Lake made it a better model, as the likely values for each don't all overlap and the likely $\sigma$ is somewhat smaller (4.93 vs 5.33 for RC0_Null).

Note that there is lots of overlap in the likely mean values for Superior, Huron, and Ontario; each of them likely has about 8 taxa/site. But Lake Erie has a much higher likely taxon richness per site… about 17 taxa. We saw this when we just looked at the data back in Fig. 3.6, and I noted that although each site tends to have more taxa in Lake Erie, we see fewer taxa across all sites than the other lakes. The conceptual model also includes Effective Fetch as a potentially important causal factor that helps

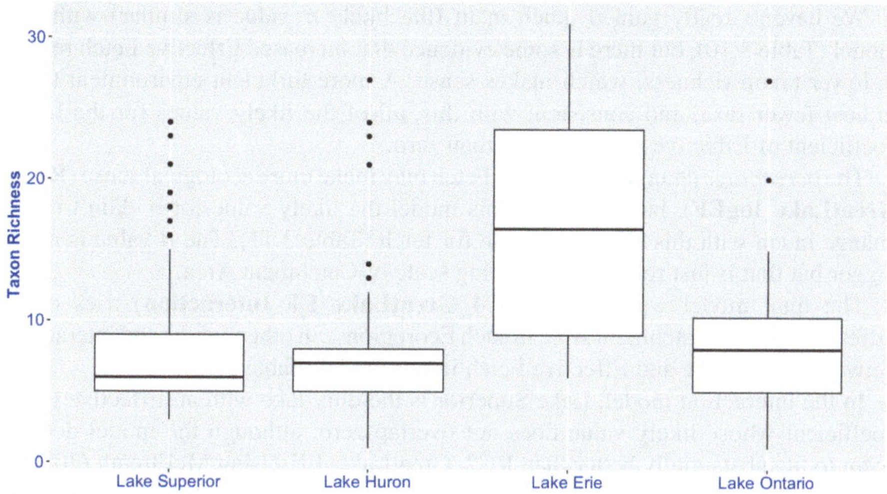


**Fig. 3.6** GLBEAST: Taxon richness of Reference sites in Lake Superior (n = 73), Lake Huron (n = 53), Lake Erie (n = 12), and Lake Ontario (n = 37)

**Table 3.8** RC0_Null—Bayesian fit of the simplest ("null") Reference Condition model of variation in taxon richness among GLBEAST ecosystems

| | Mean | Standard deviation | 5.5% | 94.5% |
|---|---|---|---|---|
| $\mu$ | 8.47 | 0.32 | 7.96 | 8.99 |
| $\sigma$ | 5.33 | 0.23 | 4.96 | 5.70 |

**Table 3.9** RC1_GreatLake—Bayesian fit of the Great Lake only Reference Condition model of variation in taxon richness among GLBEAST ecosystems

| | Mean | Standard deviation | 5.5% | 94.5% |
|---|---|---|---|---|
| $\alpha_{Superior}$ | 8.06 | 0.45 | 7.34 | 8.78 |
| $\alpha_{Huron}$ | 7.76 | 0.51 | 6.95 | 8.58 |
| $\alpha_{Erie}$ | 17.12 | 1.31 | 15.03 | 19.21 |
| $\alpha_{Ontario}$ | 8.49 | 0.74 | 7.31 | 9.68 |
| $\sigma$ | 4.93 | 0.21 | 4.59 | 5.27 |

describe the natural environment of the nearshore Great Lakes ecosystem. It seems reasonable to assume that fetch might cause at least some of the variation in the biota among Reference sites. We can use the same priors as RC1_GreatLake for each lake, try out slope values for Effective Fetch with a mean and standard deviation of 1, and then once again let the data help us decide what the most (and least) likely combinations of values are for $\alpha_{Superior}$, $\alpha_{Huron}$, $\alpha_{Erie}$, $\alpha_{Ontario}$, $\beta\_EF$ and $\sigma$.

We haven't really gained much in fit (the likely $\sigma$ value is similar) with this model (Table 3.10), but there is some evidence that increased Effective Fetch results in lower taxon richness, which makes sense. A more turbulent environment tends to host fewer taxa, and consistent with this, all of the likely values for the linear coefficient of Effective Fetch are less than zero.

The *percentage* change in Effective Fetch may make more ecological sense (**RC3_GreatLake_logEF**), but when I fit this model the likely value for $\sigma$ didn't really change much with this change in scale for fetch (Table 3.11). The $\beta$ value is much bigger but that is just reflective of the log scale of Catchment Area.

The final model we will fit (**RC4_GreatLake_EF_Interaction**) tries out a different effect of Catchment Area in each Ecoregion…in other words, and interaction between Great Lake and Effective Fetch in our model (Table 3.12).

In the interaction model, Lake Superior is the only lake with an Effective Fetch coefficient whose likely value does not overlap zero, although the model doesn't seem to fit substantially better than RC2_GreatLake_EF. Using McElreath (2020)'s method of comparing the models with WAIC, RC3_GreatLake_logEF seems to be the best model of Reference Condition (Table 3.13). Plotting actual Reference and Test site values along with the model fit shows the variation among the lakes in taxon richness with the independent, negative effect of Effective Fetch in each lake (Fig. 3.7).

**Table 3.10** **RC2_GreatLake_EF**—Bayesian fit of the great lake and effective fetch Reference Condition model of variation in taxon richness among GLBEAST ecosystems

| | Mean | Standard deviation | 5.5% | 94.5% |
|---|---|---|---|---|
| $\alpha_{Superior}$ | 8.81 | 0.48 | 8.05 | 9.57 |
| $\alpha_{Huron}$ | 6.96 | 0.53 | 6.11 | 7.82 |
| $\alpha_{Erie}$ | 17.32 | 1.27 | 15.29 | 19.35 |
| $\alpha_{Ontario}$ | 8.04 | 0.73 | 6.87 | 9.20 |
| $\beta\_EF$ | −0.063 | 0.016 | −0.088 | −0.038 |
| $\sigma$ | 4.79 | 0.20 | 4.46 | 5.12 |

**Table 3.11** **RC3_GreatLake_logEF**—Bayesian fit of the great lake and log(Effective Fetch) Reference Condition model of variation in taxon richness among GLBEAST ecosystems

| | Mean | Standard deviation | 5.5% | 94.5% |
|---|---|---|---|---|
| $\alpha_{Superior}$ | 8.94 | 0.48 | 8.17 | 9.71 |
| $\alpha_{Huron}$ | 6.68 | 0.55 | 5.80 | 7.57 |
| $\alpha_{Erie}$ | 18.03 | 1.28 | 15.98 | 20.07 |
| $\alpha_{Ontario}$ | 8.06 | 0.72 | 6.90 | 9.21 |
| $\beta\_logEF$ | −1.214 | 0.291 | −1.678 | −0.75 |
| $\sigma$ | 4.75 | 0.20 | 4.42 | 5.07 |

**Table 3.12 RC4_GreatLake_EF_Interaction**—Bayesian fit of the interaction Reference Condition model of variation in taxon richness among GLBEAST ecosystems

| | Mean | Standard deviation | 5.5% | 94.5% |
|---|---|---|---|---|
| $\alpha_{Superior}$ | 8.85 | 0.51 | 8.04 | 9.67 |
| $\beta\_EF_{Superior}$ | −1.09 | 0.36 | −1.67 | −0.51 |
| $\alpha_{Huron}$ | 7.18 | 0.70 | 6.06 | 8.30 |
| $\beta\_EF_{Huron}$ | −0.65 | 0.56 | −1.54 | 0.24 |
| $\alpha_{Erie}$ | 17.58 | 1.47 | 15.24 | 19.92 |
| $\beta\_EF_{Erie}$ | −0.44 | 0.95 | −1.96 | 1.09 |
| $\alpha_{Ontario}$ | 8.45 | 0.75 | 7.25 | 9.66 |
| $\beta\_EF_{Ontario}$ | −0.12 | 0.61 | −1.09 | 0.84 |
| $\sigma$ | 4.78 | 0.21 | 4.45 | 5.12 |

**Table 3.13** Comparing Reference Condition models of variation in species richness among GLBEAST ecosystems

| | WAIC | SE | dWAIC | dSE | pWAIC | Weight |
|---|---|---|---|---|---|---|
| RC3_GreatLake_ logEF | 1629.4 | 36.42 | 0.0 | NA | 9.4 | 0.87 |
| RC2_GreatLake_ EF | 1633.6 | 36.37 | 4.2 | 4.06 | 9.0 | 0.11 |
| RC4_GreatLake_ EF_Interaction | 1636.5 | 36.97 | 7.1 | 3.73 | 10.9 | 0.02 |
| RC1_GreatLake | 1647.3 | 37.17 | 18.0 | 7.82 | 8.1 | 0.00 |
| RC0_Null | 1681.9 | 40.04 | 52.5 | 22.96 | 3.8 | 0.00 |

## Community Composition

I've been in too many heated discussions over the years about "quantitative" versus "qualitative" (*aka* Presence/Absence) biota data. With all the uncertainty about how much diversity we are missing in sampling the biota, as well as the imprecision and inaccuracy in measuring abundance or density, I think it's safer just to look at Presence/Absence data when analyzing community variation, particularly if we have substantial variation among sites. In addition to modeling ordination scores, it is always important to look at "who is where" by producing simple tables of the particular taxa found in different groups of sites, as this will aid in interpretation of whatever modeling and bioassessment is done as well as a better understanding of the ecosystems being assessed. Note that with the GLBEAST dataset, there were several more taxa found in Test *versus* Reference sites in Lake Superior and Lake Ontario, and several more found in Reference *versus* Test sites in Lake Huron and Lake Erie (Table 3.14). This could reflect the nature of the lake, the particular Reference and Test sites sampled, or all of the above! Table perusal, which I greatly prefer to ordination biplots, can also flag the consequences of sample processing that won't

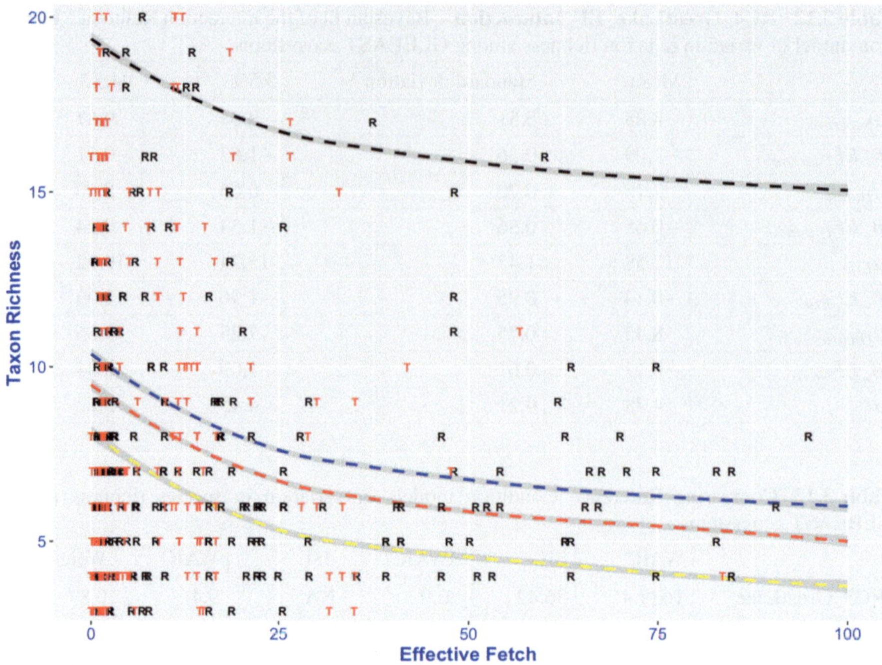


**Fig. 3.7** RC3_GreatLake_logEF fit of GLBEAST Reference sites in Lake Superior (n = 73; blue line), Lake Huron (n = 53; yellow line), Lake Erie (n = 12; black line), and Lake Ontario (n = 37; red line). Actual data from Reference (black "R") and Test (red "T") sites also shown

be evident from just looking at ordination plots. In the GLBEAST dataset there are several Trichoptera families identified but also "Trichoptera" as a separate category, presumably an ID for individuals that were early enough instars they couldn't be identified even to family.

Community composition of the GLBEAST Reference sites shows lots of variation both within and among the Great Lakes (Fig. 3.8). Note that Lake Erie, which we previously saw had the lowest $\beta$ diversity, has the least scatter in the ordination plot.

As with SimStreams, but this time using real (P/A) compositional data from the GLBEAST Reference sites, we can model the variation in composition in communities in Reference Condition. I ran the same set of models as I did for taxon richness and ended up with two, different, "best" models (Table 3.15) for NMDS1 (interaction of Great Lake and Effective Fetch) and NMDS2 (independent effects of Great Lake and Effective Fetch). A plot of the best models for NMDS1 and NMDS2, along with actual values for Reference and Test sites, is shown in Figs. 3.9 (NMDS1) and Fig. 3.10 (NMDS2).

**Table 3.14** Taxa found in Great Lakes nearshore ecosystems

| Taxon | Superior ref. | Superior test | Huron ref. | Huron test | Erie ref. | Erie test | Ontario ref. | Ontario test |
|---|---|---|---|---|---|---|---|---|
| Aeolosomatidae | | | | | X | | | X |
| Amphipoda | X | X | X | | | | | |
| Ancylidae | | X | | X | X | X | X | X |
| Anisitsiellidae | | X | X | X | | | | |
| Apataniidae | X | X | | | | | | |
| Arachnida | X | X | X | | | | X | X |
| Arrenuridae | X | X | X | X | X | X | | X |
| Asellidae | X | X | X | X | X | X | X | X |
| Aturidae | | X | X | | X | | X | X |
| Baetidae | X | | | | | | | |
| Bithyniidae | | | | | | | X | X |
| Caenidae | X | | X | X | X | X | X | X |
| Candonidae | X | X | X | X | | | X | X |
| Capniidae | | X | | | X | | | |
| Ceratopogonidae | X | X | X | X | X | | X | X |
| Chaoboridae | | X | X | X | | | X | X |
| Chironomidae | X | X | X | X | X | X | X | X |
| Coenagrionidae | | | X | | X | X | | X |
| Collembola | | X | X | | | | | |
| Corduliidae | | | | | | | | X |
| Corixidae | X | X | | | | | | |
| Crangonyctidae | | | | X | | | | |
| Cyclocyprididae | | X | X | X | | | X | X |
| Cyprididae | X | X | X | X | | | X | X |
| Dipseudopsidae | | X | X | X | X | | X | X |
| Dreissenidae | X | X | X | X | X | X | X | X |
| Dugesiidae | X | | | | X | X | X | X |
| Elmidae | X | X | X | X | | X | X | X |
| Empididae | X | X | X | | | | X | |
| Enchytraeidae | X | X | X | X | X | X | X | X |
| Ephemerellidae | | | X | X | | | | X |
| Ephemeridae | X | X | X | X | X | X | X | X |
| Ephydridae | | | | | | | | X |
| Erpobdellidae | X | X | X | | X | X | | X |
| Gammaridae | X | X | X | X | X | X | X | X |

(continued)

**Table 3.14** (continued)

| Taxon | Superior ref. | Superior test | Huron ref. | Huron test | Erie ref. | Erie test | Ontario ref. | Ontario test |
|---|---|---|---|---|---|---|---|---|
| Gastropoda | | X | X | X | | | | X |
| Glossiphoniidae | X | X | X | X | X | X | X | X |
| Gomphidae | | | X | X | | | | |
| Halacaridae | X | X | X | X | X | | X | X |
| Haliplidae | | | | X | | | | |
| Heptageniidae | X | | | | | | | |
| Hyalellidae | X | X | X | X | X | X | X | X |
| Hydridae | X | X | X | X | X | X | X | X |
| Hydrobiidae | X | X | X | X | X | X | X | X |
| Hydrodromidae | | | | X | | | | X |
| Hydropsychidae | | X | | | | | | |
| Hydroptilidae | X | X | X | X | X | X | X | X |
| Hydrozetidae | | X | X | X | X | X | | X |
| Hygrobatidae | X | X | X | X | X | X | X | X |
| Hypogastruridae | | X | | | | | | |
| Isopoda | X | X | | | | | | |
| Krendowskiidae | | X | | | | | | |
| Laversiidae | X | | | | | | | |
| Lebertiidae | X | X | X | X | X | X | X | X |
| Lepidoptera | | | | | X | X | | X |
| Lepidostomatidae | X | X | | | | | | |
| Leptoceridae | X | X | X | X | X | X | X | X |
| Limnephilidae | X | X | | | | | | |
| Limnesiidae | X | X | X | X | X | X | X | X |
| Limnocytheridae | X | X | X | X | | | X | X |
| Lumbricidae | X | | | | | | | |
| Lumbriculidae | X | X | X | | X | X | X | X |
| Lutrochidae | | | | | | | | X |
| Lymnaeidae | X | X | | X | | | X | X |
| Macrobiotidae | | X | | X | | | | X |
| Malaconothridae | | | | | | | X | |
| Metretopodidae | X | | | X | | | | |
| Mideopsidae | X | X | X | X | | | | X |
| Molannidae | X | | X | | X | | | |
| Mysidae | X | X | | | | | | |

(continued)

**Table 3.14** (continued)

| Taxon | Superior ref. | Superior test | Huron ref. | Huron test | Erie ref. | Erie test | Ontario ref. | Ontario test |
|---|---|---|---|---|---|---|---|---|
| Naididae | X | X | X | X | X | X | X | X |
| Nemertea | | | | | | | X | X |
| Nemouridae | X | | | | | | | |
| Neorhabdocoela | X | X | X | X | X | | X | X |
| Oligochaeta | | X | | | | | | |
| Ostracoda | X | X | X | X | | | | |
| Oxidae | X | X | X | X | X | | X | X |
| Phryganeidae | X | X | X | | | | | X |
| Physidae | X | X | X | | X | X | X | X |
| Pionidae | X | X | X | X | X | X | X | X |
| Piscicolidae | X | X | | | | | | |
| Pisidiidae | X | X | X | X | X | X | X | X |
| Plagiostomidae | X | X | X | X | X | X | X | X |
| Planariidae | | X | X | | X | | | X |
| Planorbidae | X | X | X | X | X | X | X | X |
| Platyhelminthes | | | X | | | | | |
| Pleuroceridae | | | | | X | X | | |
| Polycentropodidae | X | | X | X | X | | | X |
| Polycystidae | | | X | | X | | | |
| Pontoporeiidae | X | X | X | X | | X | X | |
| Psephenidae | | | | | | | | X |
| Psychodidae | | | | | | | | X |
| Pyralidae | | X | | | | | | X |
| Rhabdocoela | | X | | | | | | |
| Sabellidae | X | X | X | X | X | X | X | X |
| Sarcoptiformes | X | X | X | | | | X | X |
| Sialidae | | | X | X | | | | |
| Sisyridae | | | | | | | | X |
| Sperchontidae | | X | X | | | | X | X |
| Tardigrada | | X | X | X | X | | X | |
| Tetrastemmatidae | X | X | X | X | X | X | | X |
| Torrenticolidae | | X | X | X | X | | | X |
| Trhypochthoniidae | X | X | | | X | | | |
| Trichoptera | X | | | | | | | |
| Tricladida | | | | | | | X | X |

(continued)

**Table 3.14** (continued)

| Taxon | Superior ref. | Superior test | Huron ref. | Huron test | Erie ref. | Erie test | Ontario ref. | Ontario test |
|---|---|---|---|---|---|---|---|---|
| Trombidiformes | X | X | X | X | X | X | X | X |
| Tubificida | | X | X | | | | | |
| Turbellaria | X | X | X | X | X | X | X | X |
| Typhloplanidae | | | | X | | | | |
| Unionicolidae | | | X | X | X | X | X | X |
| Unionidae | | | | X | | | | |
| Valvatidae | X | X | X | X | X | X | X | X |
| Viviparidae | | | X | X | | X | | |
| Wettinidae | | | X | | | | | |
| **Total # Taxa** | **65** | **76** | **68** | **60** | **52** | **39** | **53** | **72** |

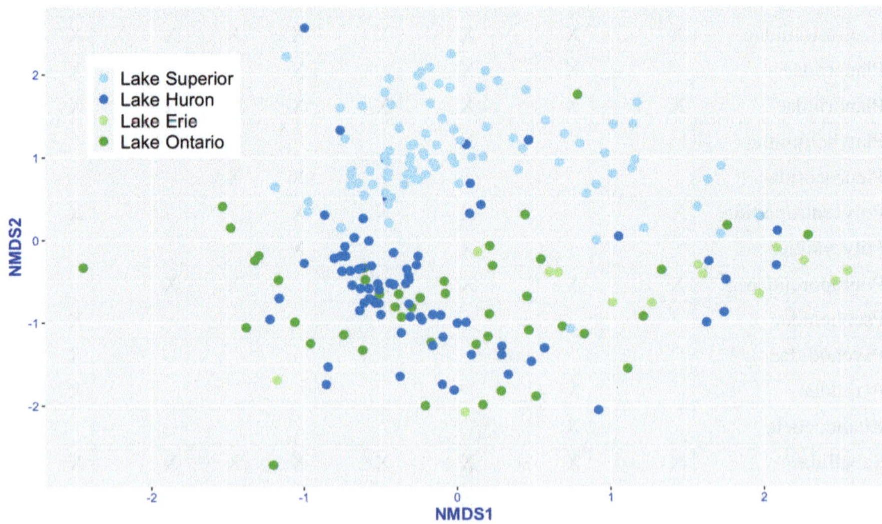


**Fig. 3.8** Community composition at GLBEAST Reference sites in Lake Superior (n = 73), Lake Huron (n = 53), Lake Erie (n = 12), and Lake Ontario (n = 37)

**Table 3.15** Bayesian fit of the Great Lake and Effective Fetch Reference Condition model of variation in ordination scores among GLBEAST ecosystems

| NMDS1 | Mean | Standard deviation | 5.5% | 94.5% |
|---|---|---|---|---|
| $\alpha_{Superior}$ | 0.089 | 0.08 | −0.12 | 0.13 |
| $\beta\_EF_{Superior}$ | −0.162 | 0.06 | −0.26 | −0.07 |
| $\alpha_{Huron}$ | −0.432 | 0.17 | −0.62 | −0.24 |
| $\beta\_EF_{Huron}$ | −0.174 | 0.10 | −0.34 | −0.01 |
| $\alpha_{Erie}$ | 2.0607 | 0.37 | 1.46 | 2.66 |
| $\beta\_EF_{Erie}$ | −1.103 | 0.41 | −1.77 | −0.43 |
| $\alpha_{Ontario}$ | −0.119 | 0.12 | −0.30 | −0.07 |
| $\beta\_EF_{Ontario}$ | −0.001 | 0.11 | −0.18 | 0.18 |
| $\sigma$ | 0.725 | 0.03 | 0.67 | 0.7 |
| **NMDS2** | | | | |
| $\alpha_{Superior}$ | 0.81 | 0.06 | 0.71 | 0.90 |
| $\alpha_{Huron}$ | −0.26 | 0.07 | −0.37 | −0.14 |
| $\alpha_{Erie}$ | −0.80 | 0.16 | −1.06 | −0.54 |
| $\alpha_{Ontario}$ | −0.66 | 0.09 | −0.81 | −0.52 |
| $\beta\_EF$ | 0.24 | 0.04 | 0.17 | 0.30 |
| $\sigma$ | 0.61 | 0.03 | 0.57 | 0.65 |

## Chapter Three Key Points

- Think about how you are **describing the biota** and don't be wed to one particular way (e.g., ordination scores, tolerance indices).
- Before modeling, **get to know the data** by looking at the variation in the biota among Reference sites grouped by one or more of the natural environment descriptors (e.g., Ecoregion, Great Lake).
- Build **Reference Condition models** that use both quantitative and categorical descriptors of the natural environment, independently and interacting, to explain variation among Reference sites in the biota.
- Choose from among the Reference Condition models you build that addresses **overfitting and generality**. Assess visualizations of the resulting model to judge the observed effects of parameters that are likely nonzero.

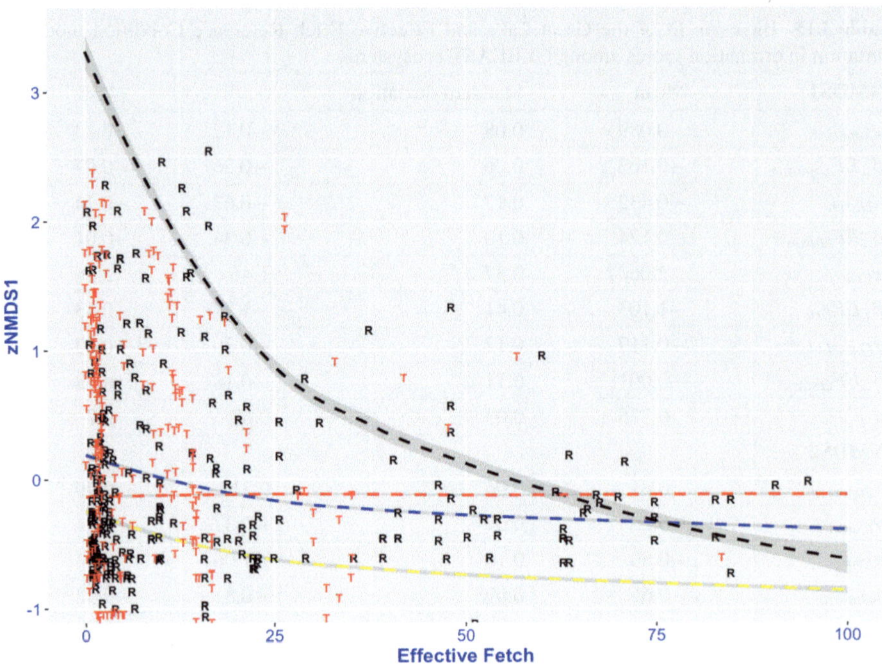


**Fig. 3.9** **GLBEAST zNMDS1**—RC4_Interaction fit of Reference sites in Lake Superior (n = 73; blue line), Lake Huron (n = 53; yellow line), Lake Erie (n = 12; black line), and Lake Ontario (n = 37; red line). Actual data from Reference (black "R") and Test (red "T") sites also shown

# Chapter Three R Scripts

## General

### 0.0 Installing Rethinking Package.R—Installs ADDIN EN.JS.CITE (McElreath 2020) rethinking and associated packages to do Bayesian model fitting

```
# First installed R 4.1.3
#Then dealt with rstan
remove.packages("rstan")
if (file.exists(".RData")) file.remove(".RData")
Sys.setenv(DOWNLOAD_STATIC_LIBV8 = 1) # only necessary for Linux without the nodejs library /
headers
# Say <no> to compiling from source
install.packages("rstan", repos = "https://cloud.r-project.org/", dependencies = TRUE)
# This was how I finally got cmdstanr installed
install.packages("cmdstanr", repos = c("https://mc-stan.org/r-packages/", getOption("repos")))
# Then ran the following line
install.packages(c("coda","mvtnorm","devtools","dagitty"))
# Then got devtools working
library(devtools)
# Then installed rethinking. say <no> to updates
devtools::install_github("rmcelreath/rethinking")
```

## SimStreams

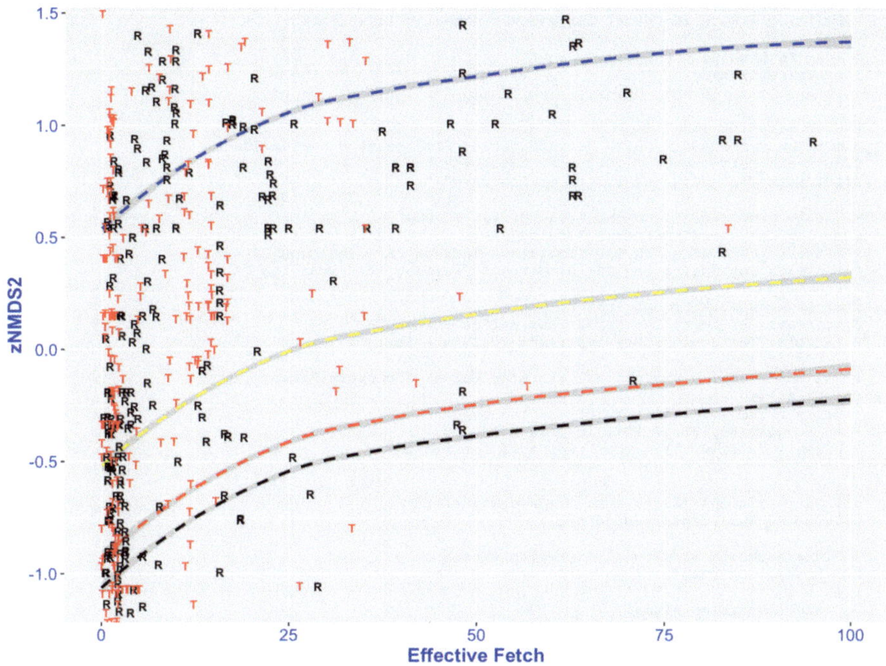


**Fig. 3.10** **GLBEAST zNMDS2**—RC4_GreatLake_logEF fit of Reference sites in Lake Superior (n = 73; blue line), Lake Huron (n = 53; yellow line), Lake Erie (n = 12; black line), and Lake Ontario (n = 37; red line). Actual data from Reference (black "R") and Test (red "T") sites also shown

## SimStreams 3_1 Boxplot of Reference Spp by Ecoregion.R—Observed species richness at Reference sites in each SimStreams Ecoregion

```
# Load necessary package(s)
library(tidyverse)
library(expss)

# Load data for plotting
SimStreams<-readRDS("./SimStreams/Datasets/SimStreams w Status.Rds")
SimStreams<-subset(SimStreams,Status=="R")

# Title, subtitle of plot, axis labels, choice of quantitative variable
output_title<-"Benthic Invertebrates in SimStreams"
qvar_title<-var_lab(SimStreams$Spp)
qvar<-SimStreams$Spp
output_subtitle<-paste(qvar_title)

# Boxplots of Species Richness by Ecoregion
ggplot(SimStreams) +
  theme(axis.text = element_text(color = "blue", size = 13)) +
  theme(axis.title = element_text(color = "blue", size = 14,face="bold")) +
  geom_boxplot(aes(Ecoregion,Spp), position = 'identity') +
  ylab(qvar_title)
```

## SimStreams 3_2 Reference Condition Model Spp from CA Ecoregion.R—Building and choosing best of models to explain variation in species richness among SimStreams Reference sites

```
# Building a Bayesian model for ecosystem health in streams

# Load rethinking package
library(rethinking)


# Load data for modelling
SimStreams<-readRDS("./SimStreams/Datasets/SimStreams w Status.Rds")
SimStreams_Ref<-subset(SimStreams,Status=="R")
SimStreams_Test<-subset(SimStreams,Status=="T")

# Create predictor Catchment Area varriables that will maintain meaningful y-intercepts
xbar_log_CA<-mean(SimStreams_Ref$log_CA)
SimStreams_Ref$delta_log_CA<-SimStreams_Ref$log_CA - xbar_log_CA
SimStreams_Test$delta_log_CA<-SimStreams_Test$log_CA - xbar_log_CA

xbar_CA<-mean(SimStreams_Ref$CA)
SimStreams_Ref$delta_CA<-SimStreams_Ref$CA - xbar_CA
SimStreams_Test$delta_CA<-SimStreams_Test$CA - xbar_CA

# Fit null model with optional starting values for optimization
flist<-alist(
  Spp ~ dnorm(mu,sigma), # This is the model
  mu ~ dnorm(50,10), # This is priors for mu
  sigma ~ dunif(0,25) # This is priors for sigma
)

EH0_Null <- quap( flist, data=SimStreams_Ref)
precis(EH0_Null,depth=2,digits=3)

# Fit a separate model with Ecoregion effect only
EH1_Ecoregion<-quap(
  alist(
    Spp ~ dnorm(mu,sigma),
    mu<-a_fit[Ecoregion],
    a_fit[Ecoregion] ~ dnorm(50,10),
    sigma ~ dunif(0,25)
  ),
  data=SimStreams_Ref)
precis(EH1_Ecoregion,depth=3,digits=3)

# Fit a model with common slope for CA and separate intercepts for each Ecoregion
EH2_Ecoregion_CA<-quap(
  alist(
    Spp ~ dnorm(mu,sigma),
    mu<-a_fit[Ecoregion] + b*(delta_CA),
    a_fit[Ecoregion] ~ dnorm(50,10),
    b ~ dnorm(1,1),
    sigma ~ dunif(0,25)
  ),
  data=SimStreams_Ref)
precis(EH2_Ecoregion_CA,depth=2,digits=3)

# Fit a model with common slope for logCA and separate intercepts for each Ecoregion
EH3_Ecoregion_logCA<-quap(
  alist(
    Spp ~ dnorm(mu,sigma),
    mu<-a_fit[Ecoregion] + b*(delta_log_CA),
        a_fit[Ecoregion] ~ dnorm(50,10),
        b ~ dnorm(1,1),
        sigma ~ dunif(0,25)
  ),
  data=SimStreams_Ref)
precis(EH3_Ecoregion_logCA,depth=2,digits=3)

# Fit a model with different slope and intercept for each group -
# interaction between age group and weight
EH4_Ecoregion_CA_Interaction<-quap(
  alist(
    Spp ~ dnorm(mu,sigma),
    mu<-a_fit[Ecoregion] + b[Ecoregion]*(delta_CA),
    a_fit[Ecoregion] ~ dnorm(50,10),
    b[Ecoregion] ~ dnorm(1,1),
    sigma ~ dunif(0,25)
```

## SimStreams 3_3 Reference Condition Best Model PredPlot Spp.R—Showing the best fitted model for species richness with the actual (simulated) SimStreams data from Reference and Test sites

```
# Building a Bayesian model for ecosystem health in streams

# Load rethinking package
library(rethinking)


# Load data for modelling
SimStreams<-readRDS("./SimStreams/Datasets/SimStreams w Status.Rds")
SimStreams_Ref<-subset(SimStreams,Status=="R")
SimStreams_Test<-subset(SimStreams,Status=="T")

# Create predictor Catchment Area variables that will maintain meaningful y-intercepts
xbar_log_CA<-mean(SimStreams_Ref$log_CA)
SimStreams_Ref$delta_log_CA<-SimStreams_Ref$log_CA - xbar_log_CA
SimStreams_Test$delta_log_CA<-SimStreams_Test$log_CA - xbar_log_CA

xbar_CA<-mean(SimStreams_Ref$CA)
SimStreams_Ref$delta_CA<-SimStreams_Ref$CA - xbar_CA
SimStreams_Test$delta_CA<-SimStreams_Test$CA - xbar_CA

# Fit null model with optional starting values for optimization
flist<-alist(
  Spp ~ dnorm(mu,sigma), # This is the model
  mu ~ dnorm(50,10), # This is priors for mu
  sigma ~ dunif(0,25) # This is priors for sigma
)

EH0_Null <- quap( flist, data=SimStreams_Ref)
precis(EH0_Null,depth=2,digits=3)

# Fit a separate model with Ecoregion effect only
EH1_Ecoregion<-quap(
  alist(
    Spp ~ dnorm(mu,sigma),
    mu<-a_fit[Ecoregion],
    a_fit[Ecoregion] ~ dnorm(50,10),
    sigma ~ dunif(0,25)
  ),
  data=SimStreams_Ref)
precis(EH1_Ecoregion,depth=3,digits=3)

# Fit a model with common slope for CA and separate intercepts for each Ecoregion
EH2_Ecoregion_CA<-quap(
  alist(
    Spp ~ dnorm(mu,sigma),
    mu<-a_fit[Ecoregion] + b*(delta_CA),
    a_fit[Ecoregion] ~ dnorm(50,10),
    b ~ dnorm(1,1),
    sigma ~ dunif(0,25)
  ),
  data=SimStreams_Ref)
precis(EH2_Ecoregion_CA,depth=2,digits=3)
```

## SimStreams 3_4 Scatter of Reference NMDS1 v NMDS2 by Ecoregion.R—Reference site community composition (ordination scores) in Highland and Valley Ecoregions

```
# Load necessary package(s)
library(tidyverse)

# Load data for plotting
SimStreams<-readRDS("./SimStreams/Datasets/SimStreams w Status.Rds")
SimStreams<-subset(SimStreams,Status=="R")


# Plot NMDS1 NMDS2 for all sites
ggplot(SimStreams,aes(zNMDS1,zNMDS2, color=Ecoregion)) +
  theme(axis.text = element_text(color = "blue", size = 10)) +
  theme(axis.title = element_text(color = "blue", size = 14,face="bold")) +
  xlab("NMDS1") +
  ylab("NMDS2") +
  theme(
    legend.position = c(.05,.95),
    legend.justification = c(0, 1),
    legend.text = element_text(size = 12),
    legend.title = element_text(size = 12,face="bold")
  ) +
  geom_point()
```

## SimStreams 3_5a Reference Condition Model zNMDS1 from CA Ecoregion.R—Building and choosing best of models to explain variation in community composition (zNMDS2) among SimStreams Reference sites

```
xbar_CA<-mean(SimStreams_Ref$CA)
SimStreams_Ref$delta_CA<-SimStreams_Ref$CA - xbar_CA
SimStreams_Test$delta_CA<-SimStreams_Test$CA - xbar_CA

# Fit null model with optional starting values for optimization
flist<-alist(
  zNMDS2 ~ dnorm(mu,sigma), # This is the model
  mu ~ dnorm(0,3), # This is priors for mu
  sigma ~ dunif(0,3) # This is priors for sigma
)

EH0_Null <- quap( flist, data=SimStreams_Ref)
precis(EH0_Null,depth=2,digits=3)

# Fit a separate model with Ecoregion effect only
EH1_Ecoregion<-quap(
  alist(
    zNMDS2 ~ dnorm(mu,sigma),
    mu<-a_fit[Ecoregion],
    a_fit[Ecoregion] ~ dnorm(0,3),
    sigma ~ dunif(0,3)
  ),
```

```
  data=SimStreams_Ref)
precis(EH1_Ecoregion,depth=3,digits=3)

# Fit a model with common slope for CA and separate intercepts for each Ecoregion
EH2_Ecoregion_CA<-quap(
  alist(
    zNMDS2 ~ dnorm(mu,sigma),
    mu<-a_fit[Ecoregion] + b*(delta_CA),
    a_fit[Ecoregion] ~ dnorm(0,3),
    b ~ dnorm(1,1),
    sigma ~ dunif(0,3)
  ),
  data=SimStreams_Ref)
precis(EH2_Ecoregion_CA,depth=2,digits=5)

# Fit a model with common slope for logCA and separate intercepts for each Ecoregion
EH3_Ecoregion_logCA<-quap(
  alist(
    zNMDS2 ~ dnorm(mu,sigma),
    mu<-a_fit[Ecoregion] + b*(delta_log_CA),
        a_fit[Ecoregion] ~ dnorm(0,3),
        b ~ dnorm(1,1),
        sigma ~ dunif(0,3)
  ),
  data=SimStreams_Ref)
precis(EH3_Ecoregion_logCA,depth=2,digits=3)

# Fit a model with different slope and intercept for each group -
# interaction between age group and weight
EH4_Ecoregion_CA_Interaction<-quap(
  alist(
    zNMDS2 ~ dnorm(mu,sigma),
    mu<-a_fit[Ecoregion] + b[Ecoregion]*(delta_CA),
```

```
# Building a Bayesian model for ecosystem health (EH) in streams

# Load rethinking package
library(rethinking)


# Load data for modelling
SimStreams<-readRDS("./SimStreams/Datasets/SimStreams w Status.Rds")
SimStreams_Ref<-subset(SimStreams,Status=="R")
SimStreams_Test<-subset(SimStreams,Status=="T")

# Create predictor Catchment Area varriables that will maintain meaningful y-intercepts
xbar_log_CA<-mean(SimStreams_Ref$log_CA)
SimStreams_Ref$delta_log_CA<-SimStreams_Ref$log_CA - xbar_log_CA
SimStreams_Test$delta_log_CA<-SimStreams_Test$log_CA - xbar_log_CA
```

## SimStreams 3_5b Reference Condition Model zNMDS2 from CA Ecoregion.R—Building and choosing best of models to explain variation in community composition (zNMDS2) among SimStreams Reference sites

```
# Building a Bayesian model for ecosystem health (EH) in streams

# Load rethinking package
library(rethinking)


# Load data for modelling
SimStreams<-readRDS("./SimStreams/Datasets/SimStreams w Status.Rds")
SimStreams_Ref<-subset(SimStreams,Status=="R")
SimStreams_Test<-subset(SimStreams,Status=="T")

# Create predictor Catchment Area varriables that will maintain meaningful y-intercepts
xbar_log_CA<-mean(SimStreams_Ref$log_CA)
SimStreams_Ref$delta_log_CA<-SimStreams_Ref$log_CA - xbar_log_CA
SimStreams_Test$delta_log_CA<-SimStreams_Test$log_CA - xbar_log_CA

xbar_CA<-mean(SimStreams_Ref$CA)
SimStreams_Ref$delta_CA<-SimStreams_Ref$CA - xbar_CA
SimStreams_Test$delta_CA<-SimStreams_Test$CA - xbar_CA

# Fit null model with optional starting values for optimization
flist<-alist(
  zNMDS2 ~ dnorm(mu,sigma), # This is the model
  mu ~ dnorm(0,3), # This is priors for mu
  sigma ~ dunif(0,3) # This is priors for sigma
)

EH0_Null <- quap( flist, data=SimStreams_Ref)
precis(EH0_Null,depth=2,digits=3)

# Fit a separate model with Ecoregion effect only
EH1_Ecoregion<-quap(
  alist(
    zNMDS2 ~ dnorm(mu,sigma),
    mu<-a_fit[Ecoregion],
    a_fit[Ecoregion] ~ dnorm(0,3),
    sigma ~ dunif(0,3)
  ),
  data=SimStreams_Ref)
precis(EH1_Ecoregion,depth=3,digits=3)

# Fit a model with common slope for CA and separate intercepts for each Ecoregion
EH2_Ecoregion_CA<-quap(
  alist(
    zNMDS2 ~ dnorm(mu,sigma),
    mu<-a_fit[Ecoregion] + b*(delta_CA),
    a_fit[Ecoregion] ~ dnorm(0,3),
    b ~ dnorm(1,1),
    sigma ~ dunif(0,3)
  ),
  data=SimStreams_Ref)
precis(EH2_Ecoregion_CA,depth=2,digits=5)

# Fit a model with common slope for logCA and separate intercepts for each Ecoregion
EH3_Ecoregion_logCA<-quap(
  alist(
    zNMDS2 ~ dnorm(mu,sigma),
    mu<-a_fit[Ecoregion] + b*(delta_log_CA),
        a_fit[Ecoregion] ~ dnorm(0,3),
        b ~ dnorm(1,1),
        sigma ~ dunif(0,3)
  ),

    data=SimStreams_Ref)
 precis(EH3_Ecoregion_logCA,depth=2,digits=5)

# Fit a model with different slope and intercept for each group -
# interaction between age group and weight
EH4_Ecoregion_CA_Interaction<-quap(
  alist(
    zNMDS2 ~ dnorm(mu,sigma),
```

## SimStreams 3_6a Reference Condition Best Model PredPlot Reference zNMDS1.R—Showing the best fitted model for community composition (zNMDS1) with the actual (simulated) SimStreams data from Reference and Test sites

```
library(rethinking)
library(tidyverse)
# From McElreath 2020 pp78-91

# 1. Prepare data for fitting
# Load data for plotting
SimStreams<-readRDS("./SimStreams/Datasets/SimStreams w Status.Rds")
SimStreams_Ref<-subset(SimStreams,Status=="R")
SimStreams_Test<-subset(SimStreams,Status=="T")

# Create predictor Catchment Area varriables that will maintain meaningful y-intercepts
SimStreams_Ref$log_CA<-log(SimStreams_Ref$CA)
SimStreams_Test$log_CA<-log(SimStreams_Test$CA)
xbar_log_CA<-mean(SimStreams_Ref$log_CA)
SimStreams_Ref$delta_log_CA<-SimStreams_Ref$log_CA - xbar_log_CA
SimStreams_Test$delta_log_CA<-SimStreams_Test$log_CA - xbar_log_CA
xbar_CA<-mean(SimStreams_Ref$CA)
SimStreams_Ref$delta_CA<-SimStreams_Ref$CA - xbar_CA
SimStreams_Test$delta_CA<-SimStreams_Test$CA - xbar_CA

# 2. Fit best model
EH2_Ecoregion_CA<-quap(
  alist(
    zNMDS1 ~ dnorm(mu,sigma),
    mu<-a_fit[Ecoregion] + b*(delta_CA),
    a_fit[Ecoregion] ~ dnorm(0,3),
    b ~ dnorm(1,1),
    sigma ~ dunif(0,3)
  ),
  data=SimStreams_Ref)
precis(EH2_Ecoregion_CA,depth=2,digits=5)

# 3 Given the best guess at the parameters, figure out likelihood of actual values of
# zNMDS1 given Ecoregion and Catchment Area

# Simulate 1000 Predicted Values for every Reference site
zNMDS1_Pred<-link(EH2_Ecoregion_CA,data=SimStreams_Ref)
# Calculate Mean Predicted Value for every Reference site
SimStreams_Ref$mu.zNMDS1_Pred<-apply(zNMDS1_Pred,2,mean)
SimStreams_Ref$zNMDS1_Resid<-SimStreams_Ref$zNMDS1 - SimStreams_Ref$mu.zNMDS1_Pred
```

```
# Simulate 1000 Predicted Value for every Test site
zNMDS1_Pred<-link(EH2_Ecoregion_CA,data=SimStreams_Test)
# Calculate Mean Predicted Value for every Test site
SimStreams_Test$mu.zNMDS1_Pred<-apply(zNMDS1_Pred,2,mean)
SimStreams_Test$zNMDS1_Resid<-SimStreams_Test$zNMDS1 - SimStreams_Test$mu.zNMDS1_Pred

SimStreams_All<-rbind(SimStreams_Ref,SimStreams_Test)

# 5. Build cdf of residual values and compare actual Test and ref sites)
Resid_cdf<-ecdf(SimStreams_Ref$zNMDS1_Resid)
SimStreams_All$lpost<-Resid_cdf(SimStreams_All$zNMDS1_Resid)
SimStreams_All$assess_post<-ifelse(SimStreams_All$lpost < 0.10,"VLOW",
       ifelse(SimStreams_All$lpost < 0.20, "LOW",
              ifelse(SimStreams_All$lpost > 0.90,"VHIGH",
                     ifelse(SimStreams_All$lpost > 0.80, "HIGH", "OK"))))
SimStreams_All$assess_post <- factor(SimStreams_All$assess_post, levels =
c("VLOW","LOW","OK","HIGH","VHIGH"))

# Create a dataset [SimStreams] grid of Catchment Area (100 to 1000ha) and Ecoregion (1,2)
values
# Ecoregion 1 = Highlands 2 = Valley
# For each of two Ecoregions, 25*25 = 625 combinations of Catchment Area and Farming in
dataset
CA.list<-seq(from=100,to=1000,length.out=100)
Ecoregion.list<-c(1,2)
SimStreams_Pred<-expand.grid(Ecoregion=Ecoregion.list,CA=CA.list)
SimStreams_Pred$delta_CA<-SimStreams_Pred$CA - xbar_CA
SimStreams_Sim<-sim(EH2_Ecoregion_CA,data=SimStreams_Pred,n=100)
SimStreams_Pred$zNMDS1_Pred<-apply(SimStreams_Sim,2,mean)

SimStreams_Pred$Ecoregion<-factor(SimStreams_Pred$Ecoregion,
                                  levels = c(2,1),
                                  labels = c("Highland","Valley"))
ggplot(SimStreams_Pred,aes(CA,zNMDS1_Pred,color=Ecoregion))+
  theme(axis.text = element_text(color = "blue", size = 12)) +
  theme(axis.title = element_text(color = "blue", size = 14,face="bold")) +
  xlab("Catchment Area (ha)") +
  ylab("zNMDS1") +
  geom_smooth(linetype="dashed",show.legend=FALSE) +
  scale_color_manual(values = c( "blue", "yellow", "black","red"))+
  geom_text(data=SimStreams_All,aes(color=Status),
          x=SimStreams_All$CA,y=SimStreams_All$zNMDS1,fontface="bold",
          show.legend=FALSE,label=SimStreams_All$Status,size=3)
```

## SimStreams 3_6b Reference Condition Best Model PredPlot Reference zNMDS2.R—Showing the best fitted model for community composition (zNMDS2) with the actual (simulated) SimStreams data from Reference and Test sites

```
library(rethinking)
library(tidyverse)
# From McElreath 2020 pp78-91

# 1. Prepare data for fitting
# Load data for plotting
SimStreams<-readRDS("./SimStreams/Datasets/SimStreams w Status.Rds")
SimStreams_Ref<-subset(SimStreams,Status=="R")
SimStreams_Test<-subset(SimStreams,Status=="T")

# Create predictor Catchment Area varriables that will maintain meaningful y-intercepts
SimStreams_Ref$log_CA<-log(SimStreams_Ref$CA)
SimStreams_Test$log_CA<-log(SimStreams_Test$CA)
xbar_log_CA<-mean(SimStreams_Ref$log_CA)
SimStreams_Ref$delta_log_CA<-SimStreams_Ref$log_CA - xbar_log_CA
SimStreams_Test$delta_log_CA<-SimStreams_Test$log_CA - xbar_log_CA
xbar_CA<-mean(SimStreams_Ref$CA)
SimStreams_Ref$delta_CA<-SimStreams_Ref$CA - xbar_CA
SimStreams_Test$delta_CA<-SimStreams_Test$CA - xbar_CA

# 2. Fit best model
EH2_Ecoregion_CA<-quap(
  alist(
    zNMDS2 ~ dnorm(mu,sigma),
    mu<-a_fit[Ecoregion] + b*(delta_CA),
    a_fit[Ecoregion] ~ dnorm(0,3),
    b ~ dnorm(1,1),
    sigma ~ dunif(0,3)
  ),
  data=SimStreams_Ref)
precis(EH2_Ecoregion_CA,depth=2,digits=5)


# 3 Given the best guess at the parameters, figure out likelihood of actual values of
# zNMDS2 given Ecoregion and Catchment Area

# Simulate 1000 Predicted Values for every Reference site
zNMDS2_Pred<-link(EH2_Ecoregion_CA,data=SimStreams_Ref)
# Calculate Mean Predicted Value for every Reference site
SimStreams_Ref$mu.zNMDS2_Pred<-apply(zNMDS2_Pred,2,mean)
SimStreams_Ref$zNMDS2_Resid<-SimStreams_Ref$zNMDS2 - SimStreams_Ref$mu.zNMDS2_Pred

# Simulate 1000 Predicted Value for every Test site
zNMDS2_Pred<-link(EH2_Ecoregion_CA,data=SimStreams_Test)
# Calculate Mean Predicted Value for every Test site
SimStreams_Test$mu.zNMDS2_Pred<-apply(zNMDS2_Pred,2,mean)
SimStreams_Test$zNMDS2_Resid<-SimStreams_Test$zNMDS2 - SimStreams_Test$mu.zNMDS2_Pred

SimStreams_All<-rbind(SimStreams_Ref,SimStreams_Test)

# 5. Build cdf of residual values and compare actual Test and ref sites)
Resid_cdf<-ecdf(SimStreams_Ref$zNMDS2_Resid)
SimStreams_All$lpost<-Resid_cdf(SimStreams_All$zNMDS2_Resid)
SimStreams_All$assess_post<-ifelse(SimStreams_All$lpost < 0.10,"VLOW",
       ifelse(SimStreams_All$lpost < 0.20, "LOW",
              ifelse(SimStreams_All$lpost > 0.90,"VHIGH",
                     ifelse(SimStreams_All$lpost > 0.80, "HIGH", "OK"))))
SimStreams_All$assess_post <- factor(SimStreams_All$assess_post, levels =
c("VLOW","LOW","OK","HIGH","VHIGH"))

# Create a dataset [SimStreams] grid of Catchment Area (100 to 1000ha) and Ecoregion (1,2)
values
# Ecoregion 1 = Highlands 2 = Valley
# For each of two Ecoregions, 25*25 = 625 combinations of Catchment Area and Farming in
dataset
CA.list<-seq(from=100,to=1000,length.out=100)
Ecoregion.list<-c(1,2)
SimStreams_Pred<-expand.grid(Ecoregion=Ecoregion.list,CA=CA.list)
SimStreams_Pred$delta_CA<-SimStreams_Pred$CA - xbar_CA
SimStreams_Sim<-sim(EH2_Ecoregion_CA,data=SimStreams_Pred,n=100)
SimStreams_Pred$zNMDS2_Pred<-apply(SimStreams_Sim,2,mean)

SimStreams_Pred$Ecoregion<-factor(SimStreams_Pred$Ecoregion,
```

## GLBEAST

# GLBEAST 3_0 Calculating Indices by Lake and Status.R

```
.# Load necessary package(s)
library(tidyverse)

# Load data
GLBEAST<-readRDS("./GLBEAST/Datasets/GLBEAST RCA2.Rds")

GLBEAST<- group_by(GLBEAST,GreatLake.x,StatusRCA)

GLBEAST_byLake_Status<-summarise(GLBEAST,across(48:188, list(sum = sum)))


# Index calculations
# Create vectors that will hold calculated indices
GLBEAST_byLake_Status$Richness<-c(rep(0,nrow(GLBEAST_byLake_Status)))
GLBEAST_byLake_Status$Abundance<-c(rep(0,nrow(GLBEAST_byLake_Status)))
GLBEAST_byLake_Status$Simpson<-c(rep(0,nrow(GLBEAST_byLake_Status)))
GLBEAST_byLake_Status$Even<-c(rep(0,nrow(GLBEAST_byLake_Status)))
GLBEAST_byLake_Status$Shannon<-c(rep(0,nrow(GLBEAST_byLake_Status)))
tax_pi<-matrix(0,nrow=nrow(GLBEAST_byLake_Status),ncol=ncol(GLBEAST_byLake_Status))
tax_pi_squared<-matrix(0,nrow=nrow(GLBEAST_byLake_Status),ncol=ncol(GLBEAST_byLake_Status))

# Loop over taxa to determine Richness and Abundance
for(row in 1:nrow(GLBEAST_byLake_Status)) {
  for(col in 3:141) {
    # Increment richness for each taxon present
    if(GLBEAST_byLake_Status[row,col]>0) {GLBEAST_byLake_Status$Richness[row]<-
GLBEAST_byLake_Status$Richness[row]+1}
    # Increment abundance by number of individuals in taxon
    GLBEAST_byLake_Status$Abundance[row]<-
as.numeric(GLBEAST_byLake_Status$Abundance[row]+GLBEAST_byLake_Status[row,col])
  }
  for(col in 3:141) {
    # Determine pi for each taxon and then build diversity index calc'ns
    tax_pi[row,col]<-
as.numeric(GLBEAST_byLake_Status[row,col]/GLBEAST_byLake_Status$Abundance[row])
    tax_pi_squared[row,col]<-tax_pi[row,col]*tax_pi[row,col]
    GLBEAST_byLake_Status$Simpson[row]<-
GLBEAST_byLake_Status$Simpson[row]+tax_pi_squared[row,col]
    if(tax_pi[row,col]>0) {GLBEAST_byLake_Status$Shannon[row]<-
GLBEAST_byLake_Status$Shannon[row] - (tax_pi[row,col] * log(tax_pi[row,col]))}
  }
  GLBEAST_byLake_Status$SimpsonDinv<-1/GLBEAST_byLake_Status$Simpson
  GLBEAST_byLake_Status$Even<-GLBEAST_byLake_Status$SimpsonDinv/GLBEAST_byLake_Status$Richness
  }

# Build and store datafile with indices
iGLBEAST_byLake_Status<-
cbind(GLBEAST_byLake_Status$GreatLake.x,GLBEAST_byLake_Status$StatusRCA,

GLBEAST_byLake_Status$Richness,GLBEAST_byLake_Status$Abundance,GLBEAST_byLake_Status$Simpson,G
LBEAST_byLake_Status$SimpsonDinv,
                              GLBEAST_byLake_Status$Even,GLBEAST_byLake_Status$Shannon,
                              GLBEAST_byLake_Status[,3:ncol(GLBEAST_byLake_Status)]
saveRDS(iGLBEAST_byLake_Status,file="./GLBEAST/Datasets/GLBEAST BenthosIndices_byLake.Rds")

# Load necessary package(s)
library(tidyverse)
library(expss)

# Load data for plotting
GLBEAST_HA<-readRDS("./GLBEAST/Datasets/GLBEAST Env.Rds")

# choice of quantitative variable
qvar_title<-var_lab(GLBEAST_HA$Fetch)

# Boxplots of qvar by Great Lake
ggplot(GLBEAST_HA) +
  theme(axis.text = element_text(color = "blue", size = 15)) +
  theme(axis.title = element_text(color = "blue", size = 14,face="bold")) +
  geom_boxplot(aes(GreatLake,Fetch), position = 'identity') +
  xlab(" ") +
# facet_wrap(~ Status) +
  ylab("Effective Fetch (km)")
```

## GLBEAST 3_1 Boxplot of Richness by Lake.R—Observed taxon richness at Reference sites in each Great Lake

```
library(rethinking)
library(tidyverse)
# From McElreath 2020 pp78-91

# Prepare data for fitting
GLBEAST<-readRDS("./GLBEAST/Datasets/GLBEAST RCA2_trim.Rds")
GLBEAST$log_Fetch<-log(1+GLBEAST$Fetch)
GLBEAST_Ref<-subset(GLBEAST,StatusRCA=="R")
GLBEAST_Test<-subset(GLBEAST,StatusRCA=="T")

# Create predictor Catchment Area variables that will maintain meaningful y-intercepts
xbar_log_Fetch<-mean(GLBEAST_Ref$log_Fetch)
GLBEAST_Ref$delta_log_Fetch<-GLBEAST_Ref$log_Fetch - xbar_log_Fetch
GLBEAST_Test$delta_log_Fetch<-GLBEAST_Test$log_Fetch - xbar_log_Fetch

xbar_Fetch<-mean(GLBEAST_Ref$Fetch)
GLBEAST_Ref$delta_Fetch<-GLBEAST_Ref$Fetch - xbar_Fetch
GLBEAST_Test$delta_Fetch<-GLBEAST_Test$Fetch - xbar_Fetch


# Fit null model with optional starting values for optimization
flist<-alist(
  Richness ~ dnorm(mu,sigma), # This is the model
  mu ~ dnorm(20,10), # This is priors for mu
  sigma ~ dunif(0,10) # This is priors for sigma
)
# startvals<-list(
#  mu=mean(GLBEAST_Ref$Richness),
#  sigma=sd(GLBEAST_Ref$Richness))
EH0_Null <- quap( flist, data=GLBEAST_Ref)
precis(EH0_Null,depth=2,digits=4)

# Fit a separate model for each group with GreatLake effect only
EH1_GreatLake<-quap(
  alist(
    Richness ~ dnorm(mu,sigma),
    mu<-a_fit[GreatLake],
    a_fit[GreatLake] ~ dnorm(20,10),
    sigma ~ dunif(0,10)
  ),
  data=GLBEAST_Ref)
precis(EH1_GreatLake,depth=2,digits=4)

# Fit a model with common slope for Fetch and separate intercepts for each group -
# independent effects of Ecoregion and Catchment Area
EH2_GreatLake_Fetch<-quap(
  alist(
    Richness ~ dnorm(mu,sigma),
    mu<-a_fit[GreatLake] + b*(delta_Fetch),
    a_fit[GreatLake] ~ dnorm(20,10),
    b ~ dnorm(1,1),
    sigma ~ dunif(0,10)
  ),
  data=GLBEAST_Ref)
precis(EH2_GreatLake_Fetch,depth=3,digits=4)

# Fit a model with common slope for Fetch and separate intercepts for each group -
# independent effects of Ecoregion and Catchment Area
EH3_GreatLake_log_Fetch<-quap(
  alist(
    Richness ~ dnorm(mu,sigma),
    mu<-a_fit[GreatLake] + b*(delta_log_Fetch),
    a_fit[GreatLake] ~ dnorm(20,10),
    b ~ dnorm(1,1),
    sigma ~ dunif(0,10)
  ),
  data=GLBEAST_Ref)
precis(EH3_GreatLake_log_Fetch,depth=3,digits=4)

# Fit a model with different slope and intercept for each group
EH4_Interaction<-quap(
  alist(
    Richness ~ dnorm(mu,sigma),
    mu<-a_fit[GreatLake] + b[GreatLake]*(delta_log_Fetch),
```

## GLBEAST 3_2 Rethinking Candidate Model Richness.R—Building and choosing best of models to explain variation in species richness among SimStreams Reference sites

```
library(rethinking)
library(tidyverse)
# From McElreath 2020 pp78-91

# 1. Prepare data for fitting
GLBEAST<-readRDS("./GLBEAST/Datasets/GLBEAST RCA2_trim.Rds")
GLBEAST$log_Fetch<-log(1+GLBEAST$Fetch)
GLBEAST_Ref<-subset(GLBEAST,StatusRCA=="R")
GLBEAST_Test<-subset(GLBEAST,StatusRCA=="T")

# Create predictor Catchment Area varriables that will maintain meaningful y-intercepts
xbar_log_Fetch<-mean(GLBEAST_Ref$log_Fetch)
GLBEAST_Ref$delta_log_Fetch<-GLBEAST_Ref$log_Fetch - xbar_log_Fetch
GLBEAST_Test$delta_log_Fetch<-GLBEAST_Test$log_Fetch - xbar_log_Fetch

xbar_Fetch<-mean(GLBEAST_Ref$Fetch)
GLBEAST_Ref$delta_Fetch<-GLBEAST_Ref$Fetch - xbar_Fetch
GLBEAST_Test$delta_Fetch<-GLBEAST_Test$Fetch - xbar_Fetch

# 2. Fit best model
EH3_GreatLake_log_Fetch<-quap(
  alist(
    Richness ~ dnorm(mu,sigma),
    mu<-a_fit[GreatLake] + b*(delta_log_Fetch),
    a_fit[GreatLake] ~ dnorm(20,10),
    b ~ dnorm(1,1),
    sigma ~ dunif(0,10)
  ),
  data=GLBEAST_Ref)
precis(EH3_GreatLake_log_Fetch,depth=2,digits=4)


# 3 Given the best guess at the parameters, figure out likelihood of actual values of
# Spp given GreatLake and Catchment Area

# Simulate 1000 Predicted Values for every Reference site
Richness_Pred<-link(EH3_GreatLake_log_Fetch,data=GLBEAST_Ref)
# Calculate Mean Predicted Value for every Reference site
GLBEAST_Ref$mu.Richness_Pred<-apply(Richness_Pred,2,mean)
GLBEAST_Ref$Richness_Resid<-GLBEAST_Ref$Richness - GLBEAST_Ref$mu.Richness_Pred

# Simulate 1000 Predicted Value for every Test site
Richness_Pred<-link(EH3_GreatLake_log_Fetch,data=GLBEAST_Test)
# Calculate Mean Predicted Value for every Test site
GLBEAST_Test$mu.Richness_Pred<-apply(Richness_Pred,2,mean)
GLBEAST_Test$Richness_Resid<-GLBEAST_Test$Richness - GLBEAST_Test$mu.Richness_Pred

GLBEAST_All<-rbind(GLBEAST_Ref,GLBEAST_Test)

# 5. Build cdf of residual values and compare actual test and ref sites)
Resid_cdf<-ecdf(GLBEAST_Ref$Richness_Resid)
GLBEAST_All$lpost<-Resid_cdf(GLBEAST_All$Richness_Resid)
GLBEAST_All$assess_post<-ifelse(GLBEAST_All$lpost < 0.10,"VLOW",
       ifelse(GLBEAST_All$lpost < 0.20, "LOW",
              ifelse(GLBEAST_All$lpost > 0.90,"VHIGH",
                     ifelse(GLBEAST_All$lpost > 0.80, "HIGH", "OK"))))
GLBEAST_All$assess_post <- factor(GLBEAST_All$assess_post, levels =
c("VLOW","LOW","OK","HIGH","VHIGH"))

# Create a dataset [GLBEAST] grid of Fetch (0 to 100km) and GreatLake (1,2,3,4) values
# GreatLake 1 = Superior 2 = Huron 3=Erie 4=Ontario
# For each of four GreatLakes, 101 Fetches from 0 to 100km
Fetch.list<-seq(from=0,to=100,length.out=101)
GreatLake.list<-c(1,2,3,4)
GLBEAST_Pred<-expand.grid(GreatLake=GreatLake.list,Fetch=Fetch.list)
```

## GLBEAST 3_3 Rethinking Best Model PredPlot Richness.R—Showing the best fitted model for taxon richness with the actual GLBEAST data from Reference and Test sites

```
# Load necessary package(s)
library(tidyverse)

# Load data
GLBEAST<-readRDS("./GLBEAST/Datasets/GLBEAST RCA2.Rds")
GLBEAST_Ref<-subset(GLBEAST,GLBEAST$StatusRCA == "R")
GLBEAST_Test<-subset(GLBEAST,GLBEAST$StatusRCA == "T")

GLBEAST_Lake<- group_by(GLBEAST_Ref,GreatLake.x)
TaxFirstCol<-49
TaxLastCol<-189
GLBEAST_byLake_Ref<-summarise(GLBEAST_Lake,across(TaxFirstCol:TaxLastCol, list(sum = sum)))
write_csv(GLBEAST_byLake_Ref,file="./GLBEAST/Datasets/GLBEAST Taxa_byLake Ref.csv")

GLBEAST_Lake<- group_by(GLBEAST_Test,GreatLake.x)
TaxFirstCol<-49
TaxLastCol<-189
GLBEAST_byLake_Test<-summarise(GLBEAST_Lake,across(TaxFirstCol:TaxLastCol, list(sum = sum)))
write_csv(GLBEAST_byLake_Test,file="./GLBEAST/Datasets/GLBEAST Taxa_byLake Test.csv")
```

## GLBEAST 3_4 Taxa by Lake Ref and Test.R—Create two spreadsheets with taxa observed at Reference and Test sites in each Great Lake

```
# Load necessary packages
library(tidyverse)
library(ggplot2)
library(RColorBrewer)

# Load and filter data
GLBEAST_Ord<-readRDS("./GLBEAST/Datasets/GLBEAST RCA2_trim.Rds")
GLBEAST_Ord<-subset(GLBEAST_Ord,StatusRCA=="R")

ggplot(data=GLBEAST_Ord,aes(x=z_PA_MDS1, y=z_PA_MDS2, color=GreatLake)) +
  scale_colour_brewer(palette = "Paired") +
  theme(axis.text = element_text(color = "blue", size = 10)) +
  theme(axis.title = element_text(color = "blue", size = 14,face="bold")) +
  xlab("NMDS1") +
  ylab("NMDS2") +
  theme(
    legend.position = c(0.05,0.9),
    legend.justification = c(0, 1),
    legend.text = element_text(size = 15),
    legend.title = element_text(size = 0,face="bold")
  ) +
  geom_point(size=3)
```

## GLBEAST 3_5 Scatterplot NMDS Ordination.R—Reference Site Community composition (ordination scores) in the Great Lakes

```
library(rethinking)
library(tidyverse)
# From McElreath 2020 pp78-91

# Prepare data for fitting
GLBEAST<-readRDS("./GLBEAST/Datasets/GLBEAST RCA2_trim.Rds")
GLBEAST$log_Fetch<-log(1+GLBEAST$Fetch)
GLBEAST_Ref<-subset(GLBEAST,StatusRCA=="R")
GLBEAST_Test<-subset(GLBEAST,StatusRCA=="T")

# Create predictor Catchment Area varriables that will maintain meaningful y-intercepts
xbar_log_Fetch<-mean(GLBEAST_Ref$log_Fetch)
GLBEAST_Ref$delta_log_Fetch<-GLBEAST_Ref$log_Fetch - xbar_log_Fetch
GLBEAST_Test$delta_log_Fetch<-GLBEAST_Test$log_Fetch - xbar_log_Fetch

xbar_Fetch<-mean(GLBEAST_Ref$Fetch)
GLBEAST_Ref$delta_Fetch<-GLBEAST_Ref$Fetch - xbar_Fetch
GLBEAST_Test$delta_Fetch<-GLBEAST_Test$Fetch - xbar_Fetch


# Fit null model with optional starting values for optimization
flist<-alist(
  z_PA_MDS2 ~ dnorm(mu,sigma), # This is the model
  mu ~ dnorm(0,3), # This is priors for mu
  sigma ~ dunif(0,3) # This is priors for sigma
)
# startvals<-list(
#  mu=mean(GLBEAST_Ref$z_PA_MDS2),
#  sigma=sd(GLBEAST_Ref$z_PA_MDS2))
EH0_Null <- quap( flist, data=GLBEAST_Ref)
precis(EH0_Null,depth=2,digits=4)

# Fit a separate model for each group with GreatLake effect only
EH1_GreatLake<-quap(
  alist(
    z_PA_MDS2 ~ dnorm(mu,sigma),
    mu<-a_fit[GreatLake],
    a_fit[GreatLake] ~ dnorm(20,10),
    sigma ~ dunif(0,10)
  ),
  data=GLBEAST_Ref)
precis(EH1_GreatLake,depth=2,digits=4)

# Fit a model with common slope for Fetch and separate intercepts for each group -
# independent effects of Ecoregion and Catchment Area
EH2_GreatLake_Fetch<-quap(
  alist(
    z_PA_MDS2 ~ dnorm(mu,sigma),
    mu<-a_fit[GreatLake] + b*(delta_Fetch),
    a_fit[GreatLake] ~ dnorm(20,10),
    b ~ dnorm(1,1),
    sigma ~ dunif(0,10)
  ),
  data=GLBEAST_Ref)
precis(EH2_GreatLake_Fetch,depth=3,digits=4)

# Fit a model with common slope for Fetch and separate intercepts for each group -
# independent effects of Ecoregion and Catchment Area
EH3_GreatLake_log_Fetch<-quap(
  alist(
    z_PA_MDS2 ~ dnorm(mu,sigma),
    mu<-a_fit[GreatLake] + b*(delta_log_Fetch),
    a_fit[GreatLake] ~ dnorm(20,10),
    b ~ dnorm(1,1),
    sigma ~ dunif(0,10)
  ),
  data=GLBEAST_Ref)
precis(EH3_GreatLake_log_Fetch,depth=3,digits=4)

# Fit a model with different slope and intercept for each group
EH4_Interaction<-quap(
  alist(
    z_PA_MDS2 ~ dnorm(mu,sigma),
    mu<-a_fit[GreatLake] + b[GreatLake]*(delta_log_Fetch),
```

## GLBEAST 3_5a Reference Condition Models NMDS1.R—Building and choosing best of models to explain variation in community composition (zNMDS1) among GLBEAST Reference sites

```
library(rethinking)
library(tidyverse)
# From McElreath 2020 pp78-91

# Prepare data for fitting
GLBEAST<-readRDS("./GLBEAST/Datasets/GLBEAST RCA2_trim.Rds")
GLBEAST$log_Fetch<-log(1+GLBEAST$Fetch)
GLBEAST_Ref<-subset(GLBEAST,StatusRCA=="R")
GLBEAST_Test<-subset(GLBEAST,StatusRCA=="T")

# Create predictor Catchment Area varriables that will maintain meaningful y-intercepts
xbar_log_Fetch<-mean(GLBEAST_Ref$log_Fetch)
GLBEAST_Ref$delta_log_Fetch<-GLBEAST_Ref$log_Fetch - xbar_log_Fetch
GLBEAST_Test$delta_log_Fetch<-GLBEAST_Test$log_Fetch - xbar_log_Fetch

xbar_Fetch<-mean(GLBEAST_Ref$Fetch)
GLBEAST_Ref$delta_Fetch<-GLBEAST_Ref$Fetch - xbar_Fetch
GLBEAST_Test$delta_Fetch<-GLBEAST_Test$Fetch - xbar_Fetch


# Fit null model with optional starting values for optimization
flist<-alist(
  z_PA_MDS2 ~ dnorm(mu,sigma), # This is the model
  mu ~ dnorm(0,3), # This is priors for mu
  sigma ~ dunif(0,3) # This is priors for sigma
)
# startvals<-list(
#  mu=mean(GLBEAST_Ref$z_PA_MDS2),
#  sigma=sd(GLBEAST_Ref$z_PA_MDS2))
EH0_Null <- quap( flist, data=GLBEAST_Ref)
precis(EH0_Null,depth=2,digits=4)

# Fit a separate model for each group with GreatLake effect only
EH1_GreatLake<-quap(
  alist(
    z_PA_MDS2 ~ dnorm(mu,sigma),
    mu<-a_fit[GreatLake],
    a_fit[GreatLake] ~ dnorm(20,10),
    sigma ~ dunif(0,10)
  ),
  data=GLBEAST_Ref)
precis(EH1_GreatLake,depth=2,digits=4)

# Fit a model with common slope for Fetch and separate intercepts for each group -
# independent effects of Ecoregion and Catchment Area
EH2_GreatLake_Fetch<-quap(
  alist(
    z_PA_MDS2 ~ dnorm(mu,sigma),
    mu<-a_fit[GreatLake] + b*(delta_Fetch),
    a_fit[GreatLake] ~ dnorm(20,10),
    b ~ dnorm(1,1),
    sigma ~ dunif(0,10)
  ),
  data=GLBEAST_Ref)
precis(EH2_GreatLake_Fetch,depth=3,digits=4)

# Fit a model with common slope for Fetch and separate intercepts for each group -
# independent effects of Ecoregion and Catchment Area
EH3_GreatLake_log_Fetch<-quap(
  alist(
    z_PA_MDS2 ~ dnorm(mu,sigma),
    mu<-a_fit[GreatLake] + b*(delta_log_Fetch),
    a_fit[GreatLake] ~ dnorm(20,10),
    b ~ dnorm(1,1),
    sigma ~ dunif(0,10)
  ),
  data=GLBEAST_Ref)
precis(EH3_GreatLake_log_Fetch,depth=3,digits=4)

# Fit a model with different slope and intercept for each group
EH4_Interaction<-quap(
  alist(
    z_PA_MDS2 ~ dnorm(mu,sigma),
    mu<-a_fit[GreatLake] + b[GreatLake]*(delta_log_Fetch),
```

## GLBEAST 3_5b Reference Condition Models NMDS2.r—Building and choosing best of models to explain variation in community composition (zNMDS2) among GLBEAST Reference sites

```
library(rethinking)
library(tidyverse)
# From McElreath 2020 pp78-91

# 1. Prepare data for fitting
GLBEAST<-readRDS("./GLBEAST/Datasets/GLBEAST RCA2_trim.Rds")
GLBEAST$log_Fetch<-log(1+GLBEAST$Fetch)
GLBEAST_Ref<-subset(GLBEAST,StatusRCA=="R")
GLBEAST_Test<-subset(GLBEAST,StatusRCA=="T")

# Create predictor Catchment Area variables that will maintain meaningful y-intercepts
xbar_log_Fetch<-mean(GLBEAST_Ref$log_Fetch)
GLBEAST_Ref$delta_log_Fetch<-GLBEAST_Ref$log_Fetch - xbar_log_Fetch
GLBEAST_Test$delta_log_Fetch<-GLBEAST_Test$log_Fetch - xbar_log_Fetch

xbar_Fetch<-mean(GLBEAST_Ref$Fetch)
GLBEAST_Ref$delta_Fetch<-GLBEAST_Ref$Fetch - xbar_Fetch
GLBEAST_Test$delta_Fetch<-GLBEAST_Test$Fetch - xbar_Fetch

# Fit a model with different slope and intercept for each group
EH4_Interaction<-quap(
  alist(
    z_PA_MDS1 ~ dnorm(mu,sigma),
    mu<-a_fit[GreatLake] + b[GreatLake]*(delta_log_Fetch),
    a_fit[GreatLake] ~ dnorm(60,20),
    b[GreatLake] ~ dnorm(1,1),
    sigma ~ dunif(0,10)
  ),
  data=GLBEAST_Ref)
precis(EH4_Interaction,depth=3,digits=4)


# 3 Given the best guess at the parameters, figure out likelihood of actual values of
# Spp given GreatLake and Catchment Area

# Simulate 1000 Predicted Values for every Reference site
z_PA_MDS1_Pred<-link(EH4_Interaction,data=GLBEAST_Ref)
# Calculate Mean Predicted Value for every Reference site
GLBEAST_Ref$mu.z_PA_MDS1_Pred<-apply(z_PA_MDS1_Pred,2,mean)
GLBEAST_Ref$z_PA_MDS1_Resid<-GLBEAST_Ref$z_PA_MDS1 - GLBEAST_Ref$mu.z_PA_MDS1_Pred

# Simulate 1000 Predicted Value for every Test site
z_PA_MDS1_Pred<-link(EH4_Interaction,data=GLBEAST_Test)
# Calculate Mean Predicted Value for every Test site
GLBEAST_Test$mu.z_PA_MDS1_Pred<-apply(z_PA_MDS1_Pred,2,mean)
GLBEAST_Test$z_PA_MDS1_Resid<-GLBEAST_Test$z_PA_MDS1 - GLBEAST_Test$mu.z_PA_MDS1_Pred

GLBEAST_All<-rbind(GLBEAST_Ref,GLBEAST_Test)

# 5. Build cdf of residual values and compare actual test and ref sites)
Resid_cdf<-ecdf(GLBEAST_Ref$z_PA_MDS1_Resid)
GLBEAST_All$lpost<-Resid_cdf(GLBEAST_All$z_PA_MDS1_Resid)
GLBEAST_All$assess_post<-ifelse(GLBEAST_All$lpost < 0.10,"VLOW",
       ifelse(GLBEAST_All$lpost < 0.20, "LOW",
              ifelse(GLBEAST_All$lpost > 0.90,"VHIGH",
                     ifelse(GLBEAST_All$lpost > 0.80, "HIGH", "OK"))))
GLBEAST_All$assess_post <- factor(GLBEAST_All$assess_post, levels =
c("VLOW","LOW","OK","HIGH","VHIGH"))

# Create a dataset [GLBEAST] grid of Fetch (0 to 100km) and GreatLake (1,2,3,4) values
# GreatLake 1 = Superior 2 = Huron 3=Erie 4=Ontario
# For each of four GreatLakes, 101 Fetches from 0 to 100km
Fetch.list<-seq(from=0,to=100,length.out=101)
GreatLake.list<-c(1,2,3,4)
GLBEAST_Pred<-expand.grid(GreatLake=GreatLake.list,Fetch=Fetch.list)
# Get log(Fetch+1) to generate predicted values
GLBEAST_Pred$log_Fetch<-log(1+GLBEAST_Pred$Fetch)
GLBEAST_Pred$delta_log_Fetch<-GLBEAST_Pred$log_Fetch - xbar_log_Fetch
# Apply model and save mean prediction for each combo of GreatLake and Fetch
GLBEAST_Sim<-sim(EH4_Interaction,data=GLBEAST_Pred,n=1000)
GLBEAST_Pred$z_PA_MDS1_Pred<-apply(GLBEAST_Sim,2,mean)

GLBEAST_Pred$GreatLake<-factor(GLBEAST_Pred$GreatLake,
                               levels = c(1,2,3,4),
```

## GLBEAST 3_6a Reference Condition Best Model PredPlot NMDS1.r—Showing the best fitted model for community composition (zNMDS1) with the actual GLBEAST Data from Reference and Test sites

```
library(rethinking)
library(tidyverse)
# From McElreath 2020 pp78-91

# 1. Prepare data for fitting
GLBEAST<-readRDS("./GLBEAST/Datasets/GLBEAST RCA2_trim.Rds")
GLBEAST$log_Fetch<-log(1+GLBEAST$Fetch)
GLBEAST_Ref<-subset(GLBEAST,StatusRCA=="R")
GLBEAST_Test<-subset(GLBEAST,StatusRCA=="T")

# Create predictor Catchment Area varriables that will maintain meaningful y-intercepts
xbar_log_Fetch<-mean(GLBEAST_Ref$log_Fetch)
GLBEAST_Ref$delta_log_Fetch<-GLBEAST_Ref$log_Fetch - xbar_log_Fetch
GLBEAST_Test$delta_log_Fetch<-GLBEAST_Test$log_Fetch - xbar_log_Fetch

xbar_Fetch<-mean(GLBEAST_Ref$Fetch)
GLBEAST_Ref$delta_Fetch<-GLBEAST_Ref$Fetch - xbar_Fetch
GLBEAST_Test$delta_Fetch<-GLBEAST_Test$Fetch - xbar_Fetch

# Fit a model with common slope for Fetch and separate intercepts for each group -
# independent effects of Ecoregion and Catchment Area
EH3_GreatLake_log_Fetch<-quap(
  alist(
    z_PA_MDS2 ~ dnorm(mu,sigma),
    mu<-a_fit[GreatLake] + b*(delta_log_Fetch),
    a_fit[GreatLake] ~ dnorm(20,10),
    b ~ dnorm(1,1),
    sigma ~ dunif(0,10)
  ),
  data=GLBEAST_Ref)
precis(EH3_GreatLake_log_Fetch,depth=3,digits=4)



# 3 Given the best guess at the parameters, figure out likelihood of actual values of
# Spp given GreatLake and Catchment Area

# Simulate 1000 Predicted Values for every Reference site
z_PA_MDS2_Pred<-link(EH3_GreatLake_log_Fetch,data=GLBEAST_Ref)
# Calculate Mean Predicted Value for every Reference site
GLBEAST_Ref$mu.z_PA_MDS2_Pred<-apply(z_PA_MDS2_Pred,2,mean)
GLBEAST_Ref$z_PA_MDS2_Resid<-GLBEAST_Ref$z_PA_MDS2 - GLBEAST_Ref$mu.z_PA_MDS2_Pred

# Simulate 1000 Predicted Value for every Test site
z_PA_MDS2_Pred<-link(EH3_GreatLake_log_Fetch,data=GLBEAST_Test)
# Calculate Mean Predicted Value for every Test site
GLBEAST_Test$mu.z_PA_MDS2_Pred<-apply(z_PA_MDS2_Pred,2,mean)
GLBEAST_Test$z_PA_MDS2_Resid<-GLBEAST_Test$z_PA_MDS2 - GLBEAST_Test$mu.z_PA_MDS2_Pred

GLBEAST_All<-rbind(GLBEAST_Ref,GLBEAST_Test)

# 5. Build cdf of residual values and compare actual test and ref sites)
Resid_cdf<-ecdf(GLBEAST_Ref$z_PA_MDS2_Resid)
GLBEAST_All$lpost<-Resid_cdf(GLBEAST_All$z_PA_MDS2_Resid)
GLBEAST_All$assess_post<-ifelse(GLBEAST_All$lpost < 0.10,"VLOW",
       ifelse(GLBEAST_All$lpost < 0.20, "LOW",
              ifelse(GLBEAST_All$lpost > 0.90,"VHIGH",
                     ifelse(GLBEAST_All$lpost > 0.80, "HIGH", "OK"))))
GLBEAST_All$assess_post <- factor(GLBEAST_All$assess_post, levels =
c("VLOW","LOW","OK","HIGH","VHIGH"))

# Create a dataset [GLBEAST] grid of Fetch (0 to 100km) and GreatLake (1,2,3,4) values
# GreatLake 1 = Superior 2 = Huron 3=Erie 4=Ontario
# For each of four GreatLakes, 101 Fetches from 0 to 100km
Fetch.list<-seq(from=0,to=100,length.out=101)
GreatLake.list<-c(1,2,3,4)
GLBEAST_Pred<-expand.grid(GreatLake=GreatLake.list,Fetch=Fetch.list)
# Get log(Fetch+1) to generate predicted values
GLBEAST_Pred$log_Fetch<-log(1+GLBEAST_Pred$Fetch)
GLBEAST_Pred$delta_log_Fetch<-GLBEAST_Pred$log_Fetch - xbar_log_Fetch
# Apply model and save mean prediction for each combo of GreatLake and Fetch
GLBEAST_Sim<-sim(EH3_GreatLake_log_Fetch,data=GLBEAST_Pred,n=1000)
GLBEAST_Pred$z_PA_MDS2_Pred<-apply(GLBEAST_Sim,2,mean)
```

### GLBEAST 3_6b Reference Condition Best Model PredPlot NMDS2.R—Showing the best fitted model for community composition (zNMDS2) with the actual GLBEAST data from Reference and Test sites

```
GLBEAST_Pred$GreatLake<-factor(GLBEAST_Pred$GreatLake,
                               levels = c(1,2,3,4),
                               labels = c("Superior","Huron","Erie","Ontario"))
GLBEAST_All$StatusRCA<-factor(GLBEAST_All$StatusRCA,
                               levels = c("R","T"))

ggplot(GLBEAST_Pred,aes(Fetch,z_PA_MDS2_Pred,color=GreatLake))+
  theme(axis.text = element_text(color = "blue", size = 12)) +
  theme(axis.title = element_text(color = "blue", size = 14,face="bold")) +
  xlab("Effective Fetch") +
  ylab("zNMDS2") +
  geom_smooth(linetype="dashed",show.legend=FALSE) +
  scale_color_manual(values = c( "blue", "yellow","black","red", "black","red")) +
  geom_text(data=GLBEAST_All,aes(color=StatusRCA),
          x=GLBEAST_All$Fetch,y=GLBEAST_All$z_PA_MDS2,fontface="bold",
          show.legend=FALSE,label=GLBEAST_All$StatusRCA,size=3)
```

## References

Anonymous: Blue Whale. Wikipedia (2024)

McElreath, R.: Statistical Rethinking: A Bayesian course with examples in R and STAN. 2nd edition. Chapman & Hall/CRC (2020)

# Chapter 4
# Bioassessment and Diagnosis of Human Activity Effects

## Assessing and Diagnosing Health

A few months ago, I visited my doctor for my annual checkup, and as she prodded and poked and noted several numbers after looking at the results of my latest blood test, I thought about the similarities between the assessment of human and ecosystem health. A doctor makes a number of observations (e.g., resting heart rate, height, weight, age) and compares these to the typical values that have been observed in a relevant population of healthy people, as well as the past data from the same person. Similarly, descriptors of the biota (e.g., species richness, community composition) in a Test ecosystem exposed to human activity can be compared to their variation among healthy ecosystems and what they were like prior to the exposure. Seeing how unusual "the patient" is relative to the observed range in healthy ecosystems (Kilgour et al. 2017) enables a judgement as to whether or not the ecosystem is healthy, and if not, why they are unhealthy (or just unusual). For example, a stream ecosystem may have the expected % shredders in the benthic invertebrate community but unusually high biodiversity. Whatever technique we use for assessment, it seems clear that the site should be reviewed to diagnose the reason for the higher than expected diversity. It may be a biodiversity hotspot, or it may be a response to increased nutrients from human activity such as farming.

After I demonstrate an approach to bioassessment… does the Test ecosystem appear to be "OK" or "normal"?…I will show and discuss the post-assessment interpretation of results including:

- What seem to be the causes of changes in the particular aspects of the biota that are unusual? (e.g., diversity in the above example)
- What is the environmental history of the assessed ecosystem in terms of exposure to human activity and natural perturbations?

Too often bioassessment programs focus on the pass-fail decision and only provide sometimes vague suggestions as to what might have caused an ecosystem to fail. Of

 
R. C. Bailey, *Rethinking Bioassessment of Freshwater Ecosystem Health*,
https://doi.org/10.1007/978-3-032-12427-2_4

course, I am happy after my annual physical to be told (at least so far!) that I am going to live for another year, but it is also good to get advice about how I could cut down on this or increase that to improve my health. In creating a bioassessment report on a Test ecosystem, we need to bring together the observed biota and its likelihood to be what we expect if the ecosystem is in Reference Condition, the exposure of the ecosystem to human activity and other stressors, and relationships from independent research as to the causal links between unusual biota descriptor values and exposure of the ecosystem.

Recall that when we use the Reference Condition Approach to bioassessment, we want to compare the biota predicted from our Reference Condition model to what we actually found at a given Test site. Here's the approach I used with the Reference Condition models I fit in Chapter Three. Using a range of specific values of the natural environment descriptors from the conceptual model (for SimStreams this would be Ecoregion and Catchment Area, for GLBEAST it would be Effective Fetch and Great Lake), simulate the distribution of the biota descriptor by plugging these values into the best Reference Condition model. The mean of these values is the ***predicted value*** of the biota descriptor in this natural environment, while the (positive or negative) ***residual*** is the difference between the observed and predicted values of the biota descriptor at a Reference or Test site. I compare the residual of each Reference and Test site to the distribution of expected values for each site from the Reference Condition model as follows:

- **Very Low**—10th percentile or less
- **Low**—10th to 20th percentile
- **OK**—20th to 80th percentile
- **High**—80th to 90th percentile
- **Very High**—90th percentile or greater

This concept is a little tricky, so let me explain it a bit more. The models aren't perfect, so even Reference sites will be somewhat different from the Reference Condition model prediction. If the model is hopelessly bad, the Reference site residuals will range from Very Low to Very High with sites in every category. If the model is great, most of the Reference sites will be close to the predicted value ("OK").

Test sites are a different matter altogether. In most situations they don't represent a population of sites that are all exposed to a similar amount of human activity…they vary in exposure, which is why we want to do the bioassessment of each in the first place! Our main interest will be in the details of individual Test site results, but we may get some general hints about the biota response to human activity by looking at any patterns in Test site residuals (see results from the example datasets below). The scripts that I have used for bioassessment of the example export spreadsheets that give the bioassessment results for individual Test sites in addition to looking at overall results for Reference and Test site groups.

For community composition (quantified with ordination scores), bioassessment of sites requires considering both NMDS1 and NMDS2 scores since together they summarize the composition. Significant deviation of the actual community composition from that expected from the Reference Condition model requires simultaneously considering both ordination scores. The categories for individual ordination scores are defined the same as the biota descriptors above (e.g. NMDS1 "Very Low" is in the 10th percentile of the distribution of values expected from the Reference Condition model:

- **Very Low**
  - NMDS1 ***or*** NMDS2 Very Low
  - NMDS1 ***and*** NMDS2 Low
- **Low**
  - NMDS1 ***or*** NMDS2 Low
- **OK**
  - NMDS1 ***and*** NMDS2 OK
- **High**
  - NMDS1 ***or*** NMDS2 High
- **Very High**
  - NMDS1 ***or*** NMDS2 Very High
  - NMDS1 ***and*** NMDS2 High

Further information about the potential causes of degradation can be gleaned by fitting a model that includes both natural and human activity environmental predictors. Such **diagnosis models** can provide some evidence of the effects of human activity and indicate whether or not human activity acts independently or interactively with the natural environment in its effects on the biota. As I show in the example datasets below, we fit these models using both Reference and Test sites since the full range of natural and human activity environmental variability will enable us to do the best job possible assessing the effects of human activity on the biota.

## *SimStreams*

A first look at the bioassessment of Test sites in SimStreams showed several with higher than expected (relative to Reference Condition model prediction) species richness at Highland Test sites and lower than expected richness at Valley Test sites (Fig. 4.1). This is a clue to what might be going on. Perhaps the human activity (farming) is having a different effect in each Ecoregion. For community composition (Fig. 4.2), the bioassessment of the Test sites shows that communities in the two Ecoregions are converging when exposed to Farming Activity (c.f. Figure 3.4).

As guided by the conceptual model for SimStreams (Fig. 2.1), I added Farming Activity to the SimStreams Reference Condition models and used both Reference and Test sites to inform the fitting. The best model for species richness in SimStreams

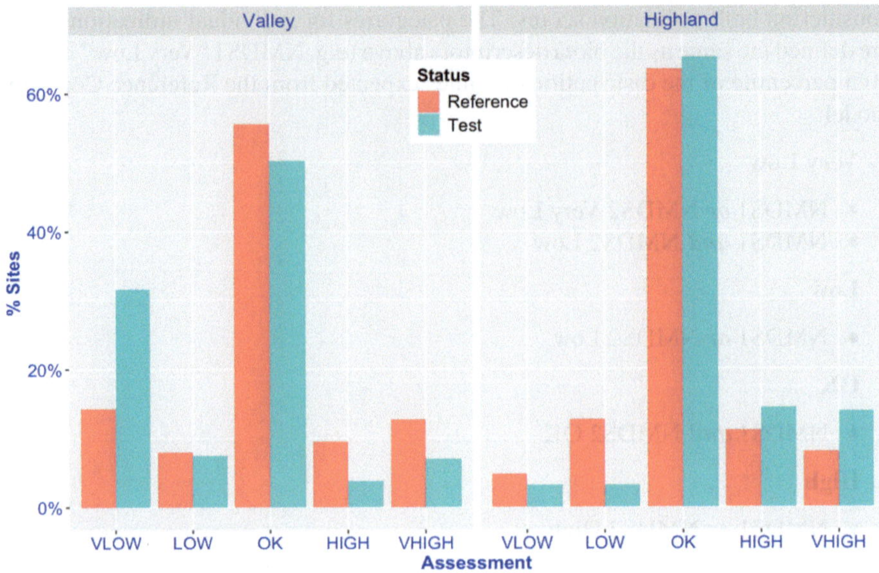


**Fig. 4.1** Comparing each Reference and Test site's actual Species Richness to that predicted by the Reference Condition model. Most Test sites far out of Reference Condition in the Valley Ecoregion have lower than expected Species Richness, while most Test sites far out of Reference Condition in the Highland Ecoregion have higher than expected Species Richness

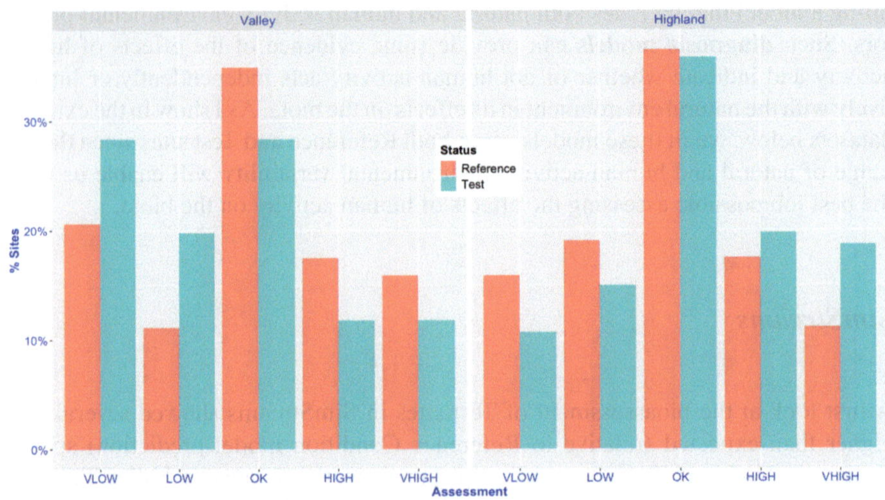


**Fig. 4.2 SimStreams Community Composition**—Comparing each Reference and Test site's actual community composition to that predicted by the Reference Condition model

(Table 4.1) included an interaction between Ecoregion and Farming Activity, showing that the effect of Farming Activity differed between the Ecoregions (Table 4.2).

Simulated data are great because you can put features into the data and see if they are actually revealed in your analysis approach. In this case, the diagnosis model shows that increased farming increases species richness in the Highland Ecoregion (increased nutrient abundance and niches perhaps?) while reducing it in the Valley Ecoregion (dominance by tolerant species and elimination of sensitive species?).

When I used this model to calculate predicted and residual values for every Reference and Test site, the pattern of Test sites that have different-than-expected species richness in each Ecoregion reflects this interaction (Fig. 4.3):

- There is higher species richness in Valley streams relative to Highland streams
- Species richness tends to increase in each Ecoregion with increasing Catchment Area
- Species richness decreases with increasing Farming Activity in Valley streams
- Species richness increases with increasing Farming Activity in Highland streams

As with species richness, the best human activity models for the ordination scores (NMDS1, NMDS2) included an interaction between Ecoregion and Farming Activity (Table 4.3).

**Table 4.1** **SimStreams Species Richness** – Choosing among human activity models (McElreath 2020). Best model is HA2_Interaction_Ecoregion, with Ecoregion, Catchment Area, and a separate slope for Farming Activity for each Ecoregion

| | WAIC | SE | dWAIC | dSE | pWAIC | weight |
|---|---|---|---|---|---|---|
| HA2_Interaction_ Ecoregion | 2563.5 | 32.39 | 0.0 | NA | 6.0 | 1 |
| HA1_Ecoregion_ CA_Farming | 2618.1 | 34.38 | 54.6 | 14.94 | 4.9 | 0 |
| HA0_Ecoregion_ CA | 2620.5 | 33.39 | 56.9 | 14.48 | 4.2 | 0 |

**Table 4.2** Bayesian fit of the HA2_Interaction_Ecoregion model of variation in species richness among SimStreams ecosystems. Note that Farming Activity has a positive effect on Highland species richness and a negative effect on Valley species richness

| | Mean | Standard Deviation | 5.5% | 94.5% |
|---|---|---|---|---|
| $\alpha_{valley}$ | 72.2 | 1.0 | 70.5 | 73.9 |
| $\alpha_{highland}$ | 45.9 | 0.6 | 45.0 | 46.8 |
| $\beta\_CA$ | 0.022 | 0.001 | 0.019 | 0.024 |
| $\beta\_FA_{valley}$ | −0.304 | 0.052 | −0.386 | −0.221 |
| $\beta\_FA_{highland}$ | 0.128 | 0.024 | 0.089 | 0.167 |
| $\sigma$ | 3.1 | 0.10 | 2.9 | 3.3 |

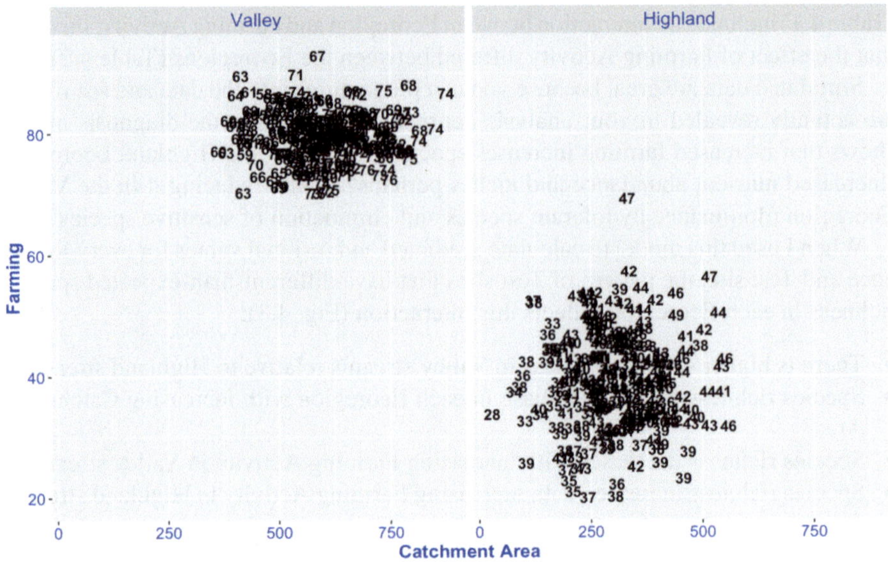


**Fig. 4.3** Species richness (plotting character) in SimStreams sites is higher in the Valley than the Highland Ecoregions, and gets higher with increasing Catchment Area in both Ecoregions. At a given Catchment Area, Species Richness gets lower with increased Farming Activity in the Valley Ecoregion and higher with increased Farming Activity in the Highland Ecoregion

**Table 4.3 SimStreams NMDS1**—Choosing among human activity models (McElreath 2020). Best model is HA2_Interaction_Ecoregion, with Ecoregion, Catchment Area, and a separate slope for Farming Activity for each Ecoregion

| NMDS1 | WAIC | SE | dWAIC | dSE | pWAIC | weight |
|---|---|---|---|---|---|---|
| HA2_Interaction_Ecoregion | 928.8 | 30.48 | 0.0 | NA | 5.8 | 0.52 |
| HA0_Ecoregion_CA | 929.9 | 30.20 | 1.1 | 4.55 | 3.8 | 0.30 |
| HA1_Ecoregion_CA_Farming | 930.8 | 30.43 | 2.0 | 4.27 | 4.9 | 0.19 |
| NMDS2 | | | | | | |
| HA2_Interaction_Ecoregion | 311.6 | 29.75 | 0.0 | NA | 6.1 | 0.99 |
| RC0_Ecoregion_CA | 323.1 | 29.18 | 11.5 | 8.53 | 3.9 | 0.00 |
| RC1_Ecoregion_CA_Farming | 323.9 | 29.51 | 12.4 | 8.40 | 4.9 | 0.00 |

The fit (Table 4.4) showed differential effects of Farming Activity on ordination scores in each Ecoregion. For both NMDS1 and NMDS2, increased farming leads to convergence of the communities in the two Ecoregions; Highland streams tend

**Table 4.4** Bayesian fit of the HA2_Interaction_Ecoregion (NMDS1, NMDS2) models of variation in ordination scores among SimStreams ecosystems

| NMDS1 | Mean | Standard Deviation | 5.5% | 94.5% |
|---|---|---|---|---|
| $\alpha_{valley}$ | 0.880 | 0.189 | 0.577 | 1.182 |
| $\alpha_{highland}$ | -0.551 | 0.102 | -0.713 | -0.389 |
| $\beta\_CA$ | 0.002 | 0.0003 | 0.001 | 0.002 |
| $\beta\_FA_{valley}$ | -0.016 | 0.009 | -0.031 | -0.001 |
| $\beta\_FA_{highland}$ | 0.001 | 0.004 | -0.006 | 0.008 |
| $\sigma$ | 0.56 | 0.02 | 0.53 | 0.59 |
| **NMDS2** | | | | |
| $\alpha_{valley}$ | 1.154 | 0.110 | 0.978 | 1.330 |
| $\alpha_{highland}$ | -0.492 | 0.059 | -0.587 | -0.398 |
| $\beta\_CA$ | 0.002 | 0.001 | 0.001 | 0.002 |
| $\beta\_FA_{valley}$ | -0.022 | 0.005 | -0.031 | -0.14 |
| $\beta\_FA_{highland}$ | 0.011 | 0.003 | 0.007 | 0.015 |
| $\sigma$ | 0.325 | 0.010 | 0.308 | 0.341 |

to have a community composition similar to Valley streams with increased farming, while with more farming Valley streams are more similar to Highland streams.

## *GLBEAST*

Bioassessment of the Test sites in GLBEAST (Fig. 4.4) showed several Lake Superior Test sites with higher than expected Taxon Richness and a similar distribution of Reference and Test site residuals in Lake Huron. Both Lake Erie and Lake Ontario had several Test sites with much lower than expected richness. This might be indicative of more substantial effects of the greater amount of human activity in the two more southern lakes, but we will reserve judgement on that until doing the diagnosis modeling.

We can look more closely at the bioassessment results through mapping (Figs. 4.5, 4.6, 4.7, 4.8, 4.9, 4.10). In Lake Superior (Fig. 4.5), several Test sites offshore of Thunder Bay and Marathon had unusually high taxon richness, perhaps due to nutrient enrichment, while a couple of Test sites in Moberley Bay and Batchawana Bay had lower than expected taxon richness. A few Test sites in Lake Huron had higher than expected taxon richness, with a couple in southern Georgian Bay with lower than expected richness (Fig. 4.6). Test sites sampled across the entire coast of Lake Erie mostly showed lower than expected taxon richness (Fig. 4.7).

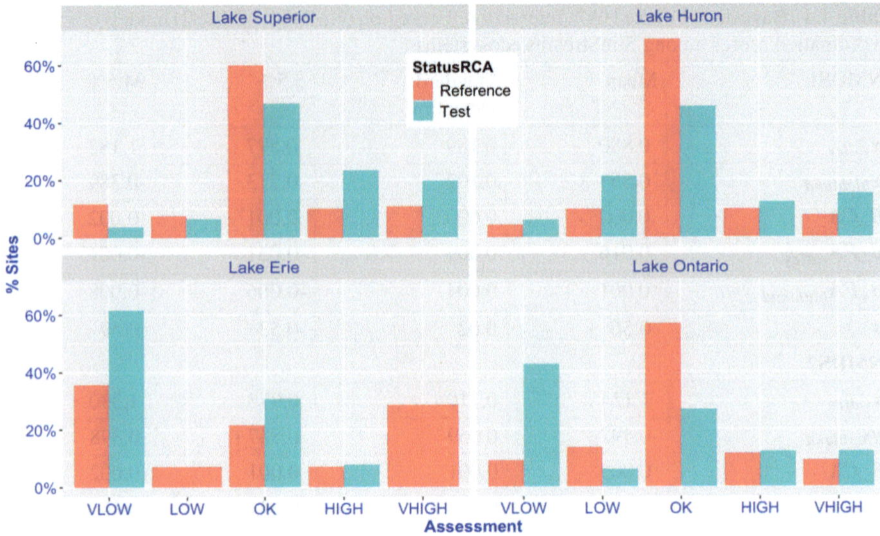


**Fig. 4.4 GLBEAST Taxon Richness** – Comparing each Reference and Test site's actual taxon richness to that predicted by the Reference Condition model (Chapter Three). Test sites tended to have lower than expected richness in Lake Erie and Lake Ontario, higher than expected richness in Lake Superior

**Fig. 4.5** Lake Superior Test sites were close to (open circles), very high (yellow circles) or very low (black circles) relative to expected Taxon Richness if in Reference Condition

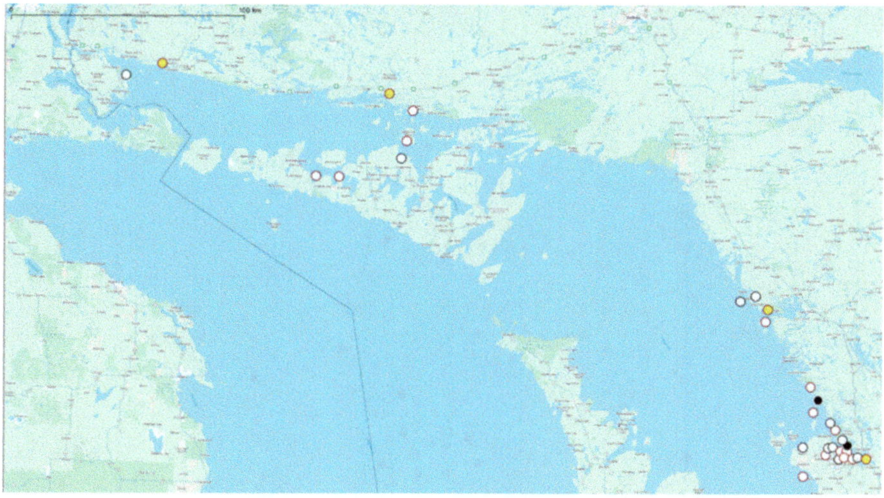

**Fig. 4.6** Lake Huron Test sites were close to (open circles), very high (yellow circles) or very low (black circles) relative to expected Taxon Richness if in Reference Condition

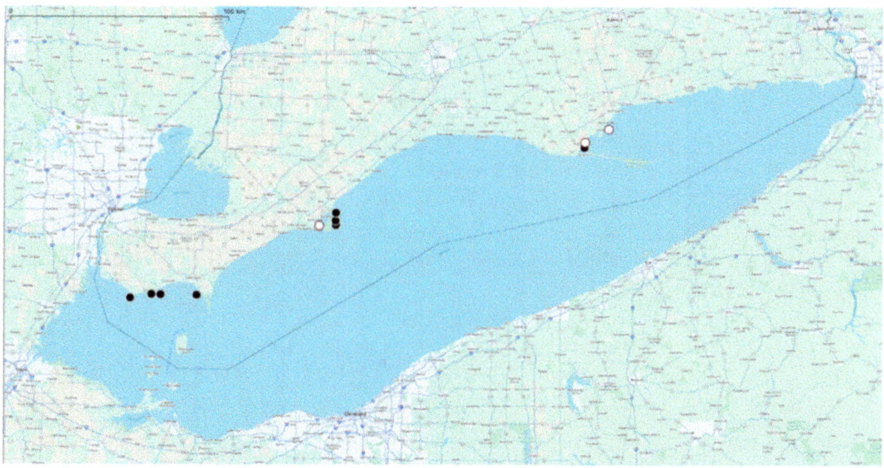

**Fig. 4.7** Lake Erie Test sites were close to (open circles), very high (yellow circles) or very low (black circles) relative to expected Taxon Richness if in Reference Condition

For Lake Ontario, most Test sites in Hamilton Harbor had lower taxon richness than expected with the exception of sites offshore of the Burlington Golf & Country Club and the Royal Botanical Gardens, which both showed higher taxon richness than expected, perhaps due to nutrient enrichment (Fig. 4.8).

In Presqu'ile Bay, unusually low taxon richness was only seen at the mouth of the Trent River in Trenton and in Picton Bay (Fig. 4.9). Otherwise, several Test sites had higher taxon richness than expected in Lake Ontario with their amount of fetch.

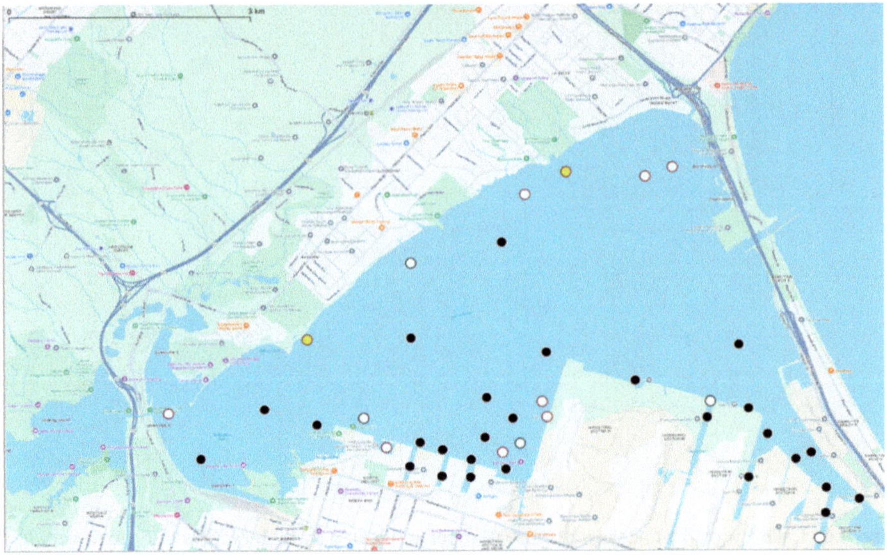

**Fig. 4.8** Test Sites in Hamilton Harbour (Lake Ontario) were close to (open circles), very high (yellow circles) or very low (black circles) relative to expected Taxon Richness if in Reference Condition

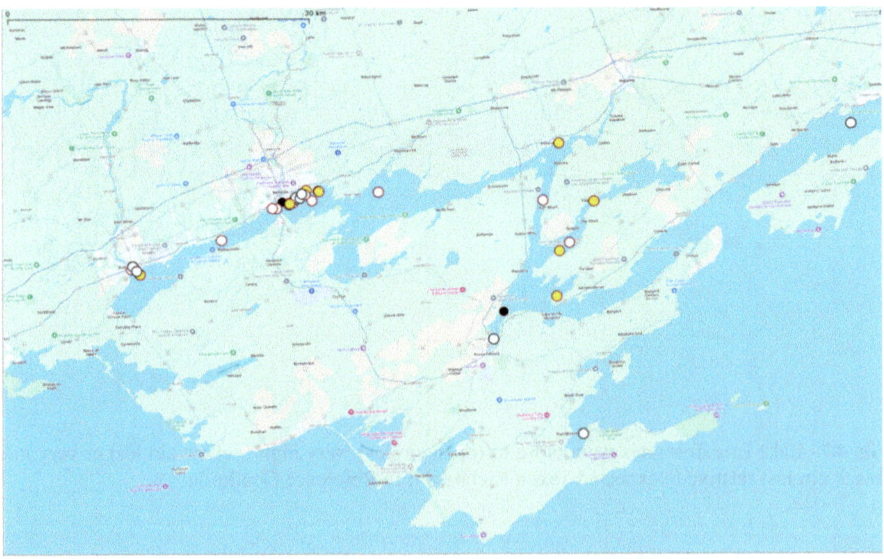

**Fig. 4.9** Test Sites in Presqu'ile Bay (Lake Ontario) were close to (open circles), very high (yellow circles) or very low (black circles) relative to expected richness if in Reference Condition

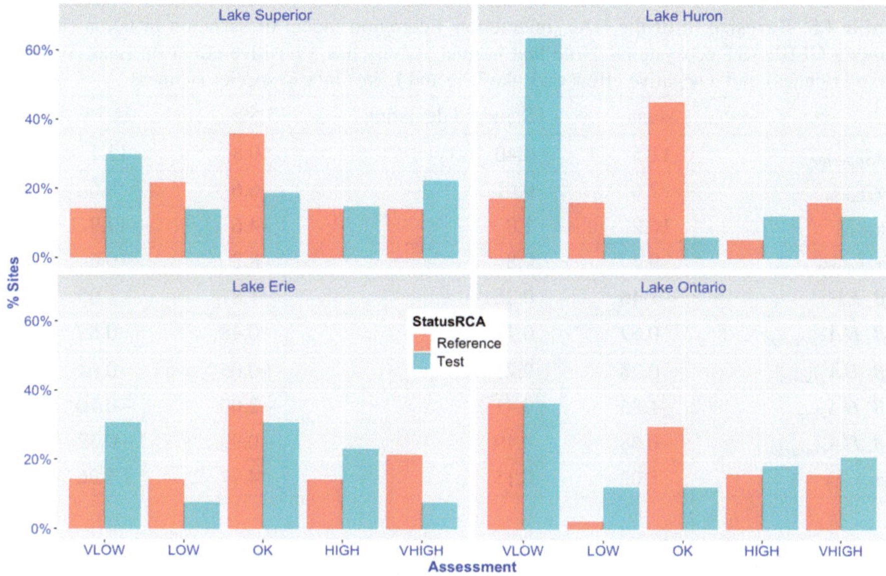


**Fig. 4.10 GLBEAST Community Composition**– Comparing each Reference and Test site's actual community composition (ordination scores) to that predicted by the Reference Condition model

Unlike the pattern observed with taxon richness, there are lots of Test sites in Lake Huron that are odd (Very Low relative to expected ordination scores) with respect to community composition (Fig. 4.10).

Looking at data from unusual sites in each lake can help in understanding potential causes of the observed effects on ecosystems. Site JB02 in Moberley Bay in Lake Superior had unusually low taxon richness ($S = 3$) and unusual community composition (Chironomidae, Naididae, Pisidiidae). Midges, naidid worms, and fingernail clams aren't unusual in Lake Superior (Appendix 1), but *just* having them is. Site 1403 in the Nort Channel of Lake Huron near Thessalon had unusually high richness ($S = 18$) and included two taxa (Haliplidae, Lymnaeidae)not seen at Reference sites. Site 1403 in Lake Erie had only three taxa (Chironomidae, Naididae, and Valvatidae), this similar to JB02 in Lake Superior in having common taxa in an uncommonly low diversity community, especially for Lake Erie. Lake Ontario site 1302 had an unusually high taxon richness ($S = 21$), but all of the taxa found were also found at Reference sites in Lake Ontario.

For GLBEAST, as depicted in the conceptual model (Fig. 2.3), the Human Activity score was added to the best Reference Condition model and it was informed by both Reference and Test sites. The best model for Taxon Richness in GLBEAST (Table 4.5) included an interaction between Great Lake and Human Activity, indicating that the effect of Human Activity differed between the lakes.

Several important conclusions arise from fitting this model. All else being equal:

**Table 4.5** Bayesian fit of the HA2_Interaction_Ecoregion model of variation in taxon richness among GLBEAST ecosystems. Note that human activity has a positive effect on Lake Superior taxon richness and a negative effect on Lake Erie and Lake Ontario species richness

| | Mean | Standard deviation | 5.5% | 94.5% |
|---|---|---|---|---|
| $\alpha_{Superior}$ | 11.5 | 0.40 | 10.8 | 12.1 |
| $\alpha_{Huron}$ | 7.6 | 0.61 | 6.6 | 8.6 |
| $\alpha_{Erie}$ | 16.2 | 1.0 | 14.6 | 17.9 |
| $\alpha_{Ontario}$ | 9.4 | 0.8 | 8.2 | 10.6 |
| $\beta\_EF$ | −1.46 | 0.26 | −1.87 | −1.05 |
| $\beta\_HA_{Superior}$ | 0.69 | 0.13 | 0.48 | 0.89 |
| $\beta\_HA_{Huron}$ | 0.28 | 0.21 | -0.05 | 0.61 |
| $\beta\_HA_{Erie}$ | −1.35 | 0.41 | −2.00 | −0.70 |
| $\beta\_HA_{Ontario}$ | −0.68 | 0.19 | −0.99 | −0.37 |
| $\sigma$ | 5.02 | 0.15 | 4.77 | 5.26 |

- The highest taxon richness per site is in Lake Erie
- The lowest taxon richness per site is in Lake Huron
- Taxon richness is lower in each Great Lake with increasing Effective Fetch
- Taxon richness increases with increasing human activity in Lake Superior
- Taxon richness decreases with increasing human activity in Lake Erie and Lake Ontario

As with taxon richness, the best human activity models for the community composition as quantified with ordination scores included an interaction between Ecoregion and Farming Activity (Table 4.6).

The fit (Table 4.7) showed differential effects of human activity on ordination scores in each Great Lake. There is evidence ($\beta\_HA_{Superior} > 0$; $\alpha_{Superior} < \alpha_{Erie}$) that greater human activity is pushing Lake Superior sites to be more like Lake Erie in composition, at least as quantified by NMDS1, whereas human activity in Lake Ontario is causing community composition to resemble that of Lake Huron sites ($\beta\_HA_{Ontario} < 0$; $\alpha_{Ontario} > \alpha_{Huron}$). For NMDS2, there is evidence that greater human activity is pushing Lake Superior sites to be more like the other lakes in composition ($\beta\_HA_{Superior} < 0$; $\alpha_{Superior} > \alpha_{Huron}$, $\alpha_{Erie}$, $\alpha_{Ontario}$), whereas human activity in Lake Ontario is causing community composition to resemble that of Lake Erie and Lake Huron sites ($\beta\_HA_{Ontarior} > 0$; $\alpha_{Ontario} < \alpha_{Huron}$, $\alpha_{Erie}$).

**Table 4.6** GLBEAST Choosing among human activity models (McElreath 2020). Best model for NMDS1 and NMDS2 is HA2_Interaction_Ecoregion, with Great Lake, Effective Fetch, and a separate slope for human activity for each Great Lake

| NMDS1 | WAIC | SE | dWAIC | dSE | pWAIC | Weight |
|---|---|---|---|---|---|---|
| HA2_Interaction_GreatLake_HAG | 1407.7 | 38.01 | 0.0 | NA | 11.5 | 1 |
| HA1_GreatLake_logEF_HAG | 1440.9 | 36.68 | 33.3 | 13.90 | 7.7 | 0 |
| HA0_GreatLake_logEF | 1441.0 | 35.94 | 33.3 | 13.42 | 6.4 | 0 |
| **NMDS2** | | | | | | |
| EH2_Interaction_GreatLake_HAG | 1139.9 | 72.21 | 0.0 | NA | 13.0 | 1 |
| EH1_GreatLake_logEF_HAG | 1158.5 | 70.21 | 18.5 | 10.58 | 10.0 | 0 |
| EH0_GreatLake_logEF | 1165.3 | 70.49 | 25.3 | 11.93 | 8.9 | 0 |

**Table 4.7** Bayesian fit of the HA2_Interaction_Ecoregion (NMDS1, NMDS2) models of variation in ordination scores among GLBEAST ecosystems

| NMDS1 | Mean | Standard deviation | 5.5% | 94.5% |
|---|---|---|---|---|
| $\alpha_{Superior}$ | 0.43 | 0.07 | 0.32 | 0.54 |
| $\alpha_{Huron}$ | −0.33 | 0.11 | −0.50 | −0.16 |
| $\alpha_{Erie}$ | 1.29 | 0.18 | 1.00 | 1.57 |
| $\alpha_{Ontario}$ | 0.03 | 0.13 | −0.19 | 0.24 |
| $\beta\_EF$ | −0.20 | 0.05 | −0.28 | −0.13 |
| $\beta\_HA_{Superior}$ | 0.11 | 0.02 | 0.07 | 0.15 |
| $\beta\_HA_{Huron}$ | 0.02 | 0.03 | −0.04 | 0.08 |
| $\beta\_HA_{Erie}$ | −0.18 | 0.41 | −2.00 | −0.70 |
| $\beta\_HA_{Ontario}$ | −0.12 | 0.03 | −0.17 | −0.06 |
| $\sigma$ | 0.87 | 0.03 | 0.83 | 0.92 |
| **NMDS2** | | | | |
| $\alpha_{Superior}$ | 0.59 | 0.05 | 0.50 | 0.68 |
| $\alpha_{Huron}$ | −0.67 | 0.08 | −0.80 | −0.54 |
| $\alpha_{Erie}$ | −0.65 | 0.14 | −0.87 | −0.43 |
| $\alpha_{Ontario}$ | −0.78 | 0.10 | −0.94 | −0.61 |
| $\beta\_EF$ | 0.15 | 0.04 | −0.09 | 0.21 |
| $\beta\_HA_{Superior}$ | −0.06 | 0.02 | −0.09 | −0.04 |
| $\beta\_HA_{Huron}$ | -0.10 | 0.03 | −0.15 | −0.06 |
| $\beta\_HA_{Erie}$ | 0.04 | 0.06 | −0.06 | 0.13 |
| $\beta\_HA_{Ontario}$ | 0.06 | 0.03 | 0.03 | 0.11 |
| $\sigma$ | 0.68 | 0.02 | 0.65 | 0.71 |

## Chapter Four Key Points

- Although the main focus of bioassessment is the evaluation of individual Test sites, patterns in **deviation between the observed and expected biota for Test sites as a group** may provide insight into the effects of human activity on the ecosystem.
- Add human activity to Reference Condition models to create **diagnosis models** that may indicate how and how much human activity affects the biota in different natural environment contexts.

## Chapter Four R Scripts

### *SimStreams*

#### SimStreams 4_1a Bioassessment Species Richness.R—Comparing Actual Species Richness at Test and Reference Sites to the Prediction from the Reference Condition Model.

```
library(rethinking)
library(tidyverse)

# 1. Prepare data for fitting
# Load data for bioassessment
SimStreams<-readRDS("./SimStreams/Datasets/SimStreams w Status.Rds")
SimStreams_Ref<-subset(SimStreams,Status=="R")
SimStreams_Test<-subset(SimStreams,Status=="T")
# Create predictor Catchment Area variables that will maintain meaningful y-intercepts
SimStreams_Ref$log_CA<-log(SimStreams_Ref$CA)
SimStreams_Test$log_CA<-log(SimStreams_Test$CA)
xbar_log_CA<-mean(SimStreams_Ref$log_CA)
SimStreams_Ref$delta_log_CA<-SimStreams_Ref$log_CA - xbar_log_CA
SimStreams_Test$delta_log_CA<-SimStreams_Test$log_CA - xbar_log_CA
xbar_CA<-mean(SimStreams_Ref$CA)
SimStreams_Ref$delta_CA<-SimStreams_Ref$CA - xbar_CA
SimStreams_Test$delta_CA<-SimStreams_Test$CA - xbar_CA

# 2 Fit best model for Spp
EH_Ecoregion_CA<-quap(
  alist(
    Spp ~ dnorm(mu,sigma),
    mu<-a_fit[Ecoregion] + b*(delta_log_CA),
    a_fit[Ecoregion] ~ dnorm(60,20),
    b ~ dnorm(1,1),
    sigma ~ dunif(0,10)
  ),
  data=SimStreams_Ref)
precis(EH_Ecoregion_CA,depth=3,digits=4)
```

```
# 2b Given the best guess at the parameters, figure out likelihood of actual values of
# Spp given Ecoregion and Catchment Area
# Simulate 1000 Predicted Values for every Reference site
Spp_Pred<-link(EH_Ecoregion_CA,data=SimStreams_Ref)
# Calculate Mean Predicted Value for every Reference site
SimStreams_Ref$mu.Spp_Pred<-apply(Spp_Pred,2,mean)
SimStreams_Ref$Spp_Resid<-SimStreams_Ref$Spp - SimStreams_Ref$mu.Spp_Pred
# Simulate 1000 Predicted Value for every Test site
Spp_Pred<-link(EH_Ecoregion_CA,data=SimStreams_Test)
# Calculate Mean Predicted Value for every Test site
SimStreams_Test$mu.Spp_Pred<-apply(Spp_Pred,2,mean)
SimStreams_Test$Spp_Resid<-SimStreams_Test$Spp - SimStreams_Test$mu.Spp_Pred

SimStreams_All_Spp<-rbind(SimStreams_Ref,SimStreams_Test)

# 3 Build cdf of residual values and compare actual test and ref sites)
Resid_cdf<-ecdf(SimStreams_Ref$Spp_Resid)
SimStreams_All_Spp$lpost<-Resid_cdf(SimStreams_All_Spp$Spp_Resid)
SimStreams_All_Spp$assess_post_Spp<-ifelse(SimStreams_All_Spp$lpost < 0.10,"VLOW",
                                   ifelse(SimStreams_All_Spp$lpost < 0.20, "LOW",
                                          ifelse(SimStreams_All_Spp$lpost > 0.90,"VHIGH",
                                                 ifelse(SimStreams_All_Spp$lpost > 0.80, "HIGH",
"OK"))))
SimStreams_All_Spp$assess_post_Spp <- factor(SimStreams_All_Spp$assess_post_Spp, levels =
c("VLOW","LOW","OK","HIGH","VHIGH"))

# 4 Plot bioassessment by Ecoregion Ref v Test

SimStreams_All_Spp$Status <- factor(SimStreams_All_Spp$Status,
                                    labels = c("Reference", "Test"))

ggplot(SimStreams_All_Spp,aes(assess_post_Spp,group=Status,fill=Status)) +
  geom_bar(aes(y=..prop..),position="dodge") +
  theme(axis.text = element_text(color = "blue", size = 13)) +
  theme(axis.title = element_text(color = "blue", size = 14,face="bold"),
        strip.text = element_text(size=14, color = "blue")) +
  theme(
    legend.position = c(.42,.95),
    legend.justification = c(0, 1),
    legend.text = element_text(size = 13),
    legend.title = element_text(size = 13,face="bold")
  ) +
  scale_y_continuous(labels = scales::percent) +
  facet_wrap(~Ecoregion) +
  xlab("Assessment") +
  ylab("% Sites")
scale_x_discrete(drop=FALSE)

write.csv(SimStreams_All_Spp,"./SimStreams/Datasets/SimStreams Bioassessment Spp.csv")
```

## SimStreams 4_1b Bioassessment Community Composition.R—Comparing Actual Community Composition (Ordination Scores) at Test and Reference Sites to the Prediction from the Reference Condition Model.

```
library(rethinking)
library(rethinking)
library(tidyverse)

# 1. Prepare data for fitting
# Load data for bioassessment
SimStreams<-readRDS("./SimStreams/Datasets/SimStreams w Status.Rds")
SimStreams_Ref<-subset(SimStreams,Status=="R")
SimStreams_Test<-subset(SimStreams,Status=="T")
# Create predictor Catchment Area variables that will maintain meaningful y-intercepts
SimStreams_Ref$log_CA<-log(SimStreams_Ref$CA)
SimStreams_Test$log_CA<-log(SimStreams_Test$CA)
xbar_log_CA<-mean(SimStreams_Ref$log_CA)
SimStreams_Ref$delta_log_CA<-SimStreams_Ref$log_CA - xbar_log_CA
SimStreams_Test$delta_log_CA<-SimStreams_Test$log_CA - xbar_log_CA
xbar_CA<-mean(SimStreams_Ref$CA)
SimStreams_Ref$delta_CA<-SimStreams_Ref$CA - xbar_CA
SimStreams_Test$delta_CA<-SimStreams_Test$CA - xbar_CA

# Fit best model for NMDS1
EH2_Ecoregion_CA_NMDS1<-quap(
  alist(
    zNMDS1 ~ dnorm(mu,sigma),
    mu<-a_fit[Ecoregion] + b*(delta_CA),
    a_fit[Ecoregion] ~ dnorm(0,3),
    b ~ dnorm(1,1),
    sigma ~ dunif(0,3)
  ),
  data=SimStreams_Ref)
precis(EH2_Ecoregion_CA_NMDS1,depth=2,digits=5)

# Given the best guess at the parameters, figure out likelihood of actual values of
# zNMDS1 given Ecoregion and Catchment Area
# Simulate 1000 Predicted Values for every Reference site
zNMDS1_Pred<-link(EH2_Ecoregion_CA_NMDS1,data=SimStreams_Ref)
# Calculate Mean Predicted Value for every Reference site
SimStreams_Ref$mu.zNMDS1_Pred<-apply(zNMDS1_Pred,2,mean)
SimStreams_Ref$zNMDS1_Resid<-SimStreams_Ref$zNMDS1 - SimStreams_Ref$mu.zNMDS1_Pred
# Simulate 1000 Predicted Value for every Test site
zNMDS1_Pred<-link(EH2_Ecoregion_CA_NMDS1,data=SimStreams_Test)
# Calculate Mean Predicted Value for every Test site
SimStreams_Test$mu.zNMDS1_Pred<-apply(zNMDS1_Pred,2,mean)
SimStreams_Test$zNMDS1_Resid<-SimStreams_Test$zNMDS1 - SimStreams_Test$mu.zNMDS1_Pred
SimStreams_All_NMDS<-rbind(SimStreams_Ref,SimStreams_Test)

# Build cdf of residual values and compare actual test and ref sites)
Resid_cdf_NMDS1<-ecdf(SimStreams_Ref$zNMDS1_Resid)
SimStreams_All_NMDS$lpost_NMDS1<-Resid_cdf_NMDS1(SimStreams_All_NMDS$zNMDS1_Resid)
SimStreams_All_NMDS$assess_post_NMDS1<-ifelse(SimStreams_All_NMDS$lpost_NMDS1 < 0.10,"VLOW",
       ifelse(SimStreams_All_NMDS$lpost_NMDS1 < 0.20, "LOW",
              ifelse(SimStreams_All_NMDS$lpost_NMDS1 > 0.90,"VHIGH",
                     ifelse(SimStreams_All_NMDS$lpost_NMDS1 > 0.80, "HIGH", "OK"))))
SimStreams_All_NMDS$assess_post_NMDS1 <- factor(SimStreams_All_NMDS$assess_post_NMDS1, levels =
c("VLOW","LOW","OK","HIGH","VHIGH"))

SimStreams_Ref<-subset(SimStreams_All_NMDS,Status=="R")
SimStreams_Test<-subset(SimStreams_All_NMDS,Status=="T")

# Fit best model for NMDS2
EH2_Ecoregion_CA_NMDS2<-quap(
  alist(
    zNMDS2 ~ dnorm(mu,sigma),
    mu<-a_fit[Ecoregion] + b*(delta_CA),
    a_fit[Ecoregion] ~ dnorm(0,3),
    b ~ dnorm(1,1),
    sigma ~ dunif(0,3)
  ),
  data=SimStreams_Ref)
precis(EH2_Ecoregion_CA_NMDS2,depth=2,digits=5)
```

```
# Given the best guess at the parameters, figure out likelihood of actual values of
# zNMDS2 given Ecoregion and Catchment Area
# Simulate 1000 Predicted Values for every Reference site
zNMDS2_Pred<-link(EH2_Ecoregion_CA_NMDS2,data=SimStreams_Ref)
# Calculate Mean Predicted Value for every Reference site
SimStreams_Ref$mu.zNMDS2_Pred<-apply(zNMDS2_Pred,2,mean)
SimStreams_Ref$zNMDS2_Resid<-SimStreams_Ref$zNMDS2 - SimStreams_Ref$mu.zNMDS2_Pred
# Simulate 1000 Predicted Value for every Test site
zNMDS2_Pred<-link(EH2_Ecoregion_CA_NMDS2,data=SimStreams_Test)
# Calculate Mean Predicted Value for every Test site
SimStreams_Test$mu.zNMDS2_Pred<-apply(zNMDS2_Pred,2,mean)
SimStreams_Test$zNMDS2_Resid<-SimStreams_Test$zNMDS2 - SimStreams_Test$mu.zNMDS2_Pred

SimStreams_All_NMDS<-rbind(SimStreams_Ref,SimStreams_Test)

# Build cdf of residual values and compare actual test and ref sites)
Resid_cdf_NMDS2<-ecdf(SimStreams_Ref$zNMDS2_Resid)
SimStreams_All_NMDS$lpost_NMDS2<-Resid_cdf_NMDS2(SimStreams_All_NMDS$zNMDS2_Resid)
SimStreams_All_NMDS$assess_post_NMDS2<-ifelse(SimStreams_All_NMDS$lpost_NMDS2 < 0.10,"VLOW",
                                    ifelse(SimStreams_All_NMDS$lpost_NMDS2 < 0.20, "LOW",
                                           ifelse(SimStreams_All_NMDS$lpost_NMDS2 > 0.90,"VHIGH",
                                                  ifelse(SimStreams_All_NMDS$lpost_NMDS2 > 0.80,
"HIGH", "OK"))))
SimStreams_All_NMDS$assess_post_NMDS2 <- factor(SimStreams_All_NMDS$assess_post_NMDS2, levels =
c("VLOW","LOW","OK","HIGH","VHIGH"))



# Do combined assessment of NMDS1 NMDS2
SimStreams_All_NMDS$assess_post_Community<-ifelse((SimStreams_All_NMDS$assess_post_NMDS1 ==
"VLOW" | SimStreams_All_NMDS$assess_post_NMDS2 == "VLOW"),"VDIF",
                                                  ifelse((SimStreams_All_NMDS$assess_post_NMDS1 ==
"LOW" & SimStreams_All_NMDS$assess_post_NMDS2 == "LOW"),"VDIF",
                                                  ifelse((SimStreams_All_NMDS$assess_post_NMDS1 ==
"LOW" | SimStreams_All_NMDS$assess_post_NMDS2 == "LOW"),"DIF",
                                                  ifelse((SimStreams_All_NMDS$assess_post_NMDS1 ==
"OK" & SimStreams_All_NMDS$assess_post_NMDS2 == "OK"),"OK",
                                                  ifelse((SimStreams_All_NMDS$assess_post_NMDS1 ==
"HIGH" & SimStreams_All_NMDS$assess_post_NMDS2 == "HIGH"),"VDIF",
                                                 ifelse((SimStreams_All_NMDS$assess_post_NMDS1 ==
"HIGH" | SimStreams_All_NMDS$assess_post_NMDS2 == "HIGH"),"DIF",
                                                  ifelse((SimStreams_All_NMDS$assess_post_NMDS1 ==
"VHIGH" | SimStreams_All_NMDS$assess_post_NMDS2 == "VHIGH"),"VDIF","ND")))))))
SimStreams_All_NMDS$assess_post_Community <- factor(SimStreams_All_NMDS$assess_post_Community,
levels = c("OK","DIF","VDIF"))

SimStreams_All_NMDS$Status <- factor(SimStreams_All_NMDS$Status,
                                labels = c("Reference", "Test"))

# Do combined assessment of NMDS1 NMDS2
ggplot(SimStreams_All_NMDS,aes(assess_post_Community,group=Status,fill=Status)) +
  geom_bar(aes(y=..prop..),position="dodge") +
  theme(axis.text = element_text(color = "blue", size = 13)) +
  theme(axis.title = element_text(color = "blue", size = 14,face="bold"),
  strip.text = element_text(size=14, color = "blue")) +
  theme(
    legend.position = c(.42,.55),
    legend.justification = c(0, 1),
    legend.text = element_text(size = 13),
    legend.title = element_text(size = 13,face="bold")
  ) +
   scale_y_continuous(labels = scales::percent) +
  facet_wrap(~Ecoregion) +
  xlab("Assessment") +
  ylab("% Sites")
scale_x_discrete(drop=FALSE)

write.csv(SimStreams_All,"./SimStreams/Datasets/SimStreams Bioassessment NMDS.csv")
```

## SimStreams 4_2 Diagnosis Model Spp from CA Ecoregion Farming.R—Finding the Best Model that Uses Both the Natural and Human Activity Environments to Explain Variation in Species Richness Among Reference and Test Sites.

```
# Building a Bayesian model for ecosystem health (EH) in streams

# Load rethinking package
library(rethinking)

# Load data for modelling
SimStreams_All<-readRDS("./SimStreams/Datasets/SimStreams w Status.Rds")

# Create predictor Catchment Area variables that will maintain meaningful y-intercepts
xbar_CA<-mean(SimStreams_All$CA)
SimStreams_All$delta_CA<-SimStreams_All$CA - xbar_CA
xbar_Farming<-mean(SimStreams_All$Farming)
SimStreams_All$delta_Farming<-SimStreams_All$Farming - xbar_Farming

# Fit a model with common slope for CA and Farming and separate intercepts for each Ecoregion
EH0_Ecoregion_CA<-quap(
  alist(
    Spp ~ dnorm(mu,sigma),
    mu<-a_fit[Ecoregion] + b_CA*(delta_CA),
    a_fit[Ecoregion] ~ dnorm(50,10),
    b_CA ~ dnorm(1,1),
    sigma ~ dunif(0,25)
  ),
  data=SimStreams_All)
precis(EH0_Ecoregion_CA,depth=2,digits=4)

# Fit a model with common slope for CA and Farming and separate intercepts for each Ecoregion
EH1_Ecoregion_CA_Farming<-quap(
  alist(
    Spp ~ dnorm(mu,sigma),
    mu<-a_fit[Ecoregion] + b_CA*(delta_CA) + b_Farming*(delta_Farming),
    a_fit[Ecoregion] ~ dnorm(50,10),
    b_CA ~ dnorm(1,1),
    b_Farming ~ dnorm(1,1),
    sigma ~ dunif(0,25)
  ),
  data=SimStreams_All)
precis(EH1_Ecoregion_CA_Farming,depth=2,digits=4)

# Fit a model with different slope of Farming for each group -
# interaction between age group and weight
EH2_Interaction_Ecoregion<-quap(
  alist(
    Spp ~ dnorm(mu,sigma),
    mu<-a_fit[Ecoregion] + b_CA*(delta_CA) + b_Farming[Ecoregion]*(delta_Farming),
    a_fit[Ecoregion] ~ dnorm(50,10),
    b_CA ~ dnorm(1,1),
    b_Farming[Ecoregion] ~ dnorm(1,1),
    sigma ~ dunif(0,25)
  ),
  data=SimStreams_All)
precis(EH2_Interaction_Ecoregion,depth=3,digits=4)

# Now compare the models - smallest WAIC wins!
compare(EH0_Ecoregion_CA,EH1_Ecoregion_CA_Farming,EH2_Interaction_Ecoregion,func=WAIC)
plot(compare(EH0_Ecoregion_CA,EH1_Ecoregion_CA_Farming,EH2_Interaction_Ecoregion))
```

## SimStreams 4_3 Diagnosis Best Model PredPlot Farming Spp.R—Showing Model Fit of Variation in Species Richness with the Natural and Human Activity Environments.

```
library(rethinking)
library(tidyverse)
# From McElreath 2020 pp78-91

# Load data for modelling
SimStreams_All<-readRDS("./SimStreams/Datasets/SimStreams w Status.Rds")

# Create predictor Catchment Area variables that will maintain meaningful y-intercepts
xbar_CA<-mean(SimStreams_All$CA)
SimStreams_All$delta_CA<-SimStreams_All$CA - xbar_CA
xbar_Farming<-mean(SimStreams_All$Farming)
SimStreams_All$delta_Farming<-SimStreams_All$Farming - xbar_Farming


# 2. Fit best model
EH2_Interaction_Ecoregion<-quap(
  alist(
    Spp ~ dnorm(mu,sigma),
    mu<-a_fit[Ecoregion] + b_CA*(delta_CA) + b_Farming[Ecoregion]*(delta_Farming),
    a_fit[Ecoregion] ~ dnorm(50,10),
    b_CA ~ dnorm(1,1),
    b_Farming[Ecoregion] ~ dnorm(1,1),
    sigma ~ dunif(0,25)
  ),
  data=SimStreams_All)
precis(EH2_Interaction_Ecoregion,depth=3,digits=4)

# 3 Given the best guess at the parameters, figure out likelihood of actual values of
# Spp given Ecoregion, Catchment Area, and Farming

# Simulate 1000 Predicted Values for every site
Spp_Pred<-link(EH2_Interaction_Ecoregion,data=SimStreams_All)
# Calculate Mean Predicted Value for every Reference site
SimStreams_All$mu.Spp_Pred<-apply(Spp_Pred,2,mean)
SimStreams_All$Spp_Resid<-SimStreams_All$Spp - SimStreams_All$mu.Spp_Pred

# Create a dataset [SimStreams] grid of Catchment Area (100 to 1000ha) and Ecoregion (1,2) values
# Ecoregion 1 = Highlands 2 = Valley
# For each of two Ecoregions, 25*25 = 625 combinations of Catchment Area and Farming
CA.list<-seq(from=1,to=1000,length.out=25)
Ecoregion.list<-c(1,2)
Farming.list<-seq(from=1,to=100,length.out=25)
SimStreams_Pred<-expand.grid(Ecoregion=Ecoregion.list,CA=CA.list,Farming=Farming.list)
SimStreams_Pred$delta_CA<-SimStreams_Pred$CA - xbar_CA
SimStreams_Pred$delta_Farming<-SimStreams_Pred$Farming - xbar_Farming
SimStreams_Sim<-sim(EH2_Interaction_Ecoregion,data=SimStreams_Pred,n=100)
SimStreams_Pred$Spp_Pred<-apply(SimStreams_Sim,2,mean)
```

```
SimStreams_Pred$Ecoregion<-factor(SimStreams_Pred$Ecoregion,
                                  levels = c(2,1),
                                  labels = c("Highland","Valley"))
# First plot fitted model for CA
ggplot(SimStreams_Pred,aes(CA,Spp_Pred,color=Ecoregion))+
  theme(axis.text = element_text(color = "blue", size = 12)) +
  theme(axis.title = element_text(color = "blue", size = 14,face="bold")) +
  xlab("Catchment Area") +
  ylab("Species Richness") +
  geom_smooth(linetype="dashed",show.legend=FALSE) +
  scale_color_manual(values = c( "blue", "yellow", "black","red"))

# Then plot fitted model for Farming
ggplot(SimStreams_Pred,aes(Farming,Spp_Pred,color=Ecoregion))+
  theme(axis.text = element_text(color = "blue", size = 12)) +
  theme(axis.title = element_text(color = "blue", size = 14,face="bold")) +
  xlab("Farming") +
  ylab("Species Richness") +
  geom_smooth(linetype="dashed",show.legend=FALSE) +
  scale_color_manual(values = c( "blue", "yellow", "black","red"))
# Then plot Farming v Catchment Area
ggplot(SimStreams_All,aes(CA,Farming))+
  theme(axis.text = element_text(color = "blue", size = 12)) +
  theme(axis.title = element_text(color = "blue", size = 14,face="bold"),
  strip.text = element_text(size=14, color = "blue")) +
  xlab("Catchment Area") +
  ylab("Farming") +
  geom_point(show.legend=FALSE,size=-1) +
  scale_color_manual(values = c("black","red", "blue", "yellow" ))+
  geom_text(data=SimStreams_All,
            x=SimStreams_All$CA,y=SimStreams_All$Farming,fontface="bold",
            show.legend=FALSE,label=SimStreams_All$Spp,size=4) +
  facet_wrap(~Ecoregion)
```

## SimStreams 4_4a Diagnosis Model NMDS1 from CA Ecoregion Farming.R—Finding the Best Model that Uses Both the Natural and Human Activity Environments to Explain Variation in NMDS1 Among Reference and Test Sites.

```
# Building a Bayesian model for ecosystem health (EH) in streams

# Load rethinking package
library(rethinking)

# Load data for modelling
SimStreams_All<-readRDS("./SimStreams/Datasets/SimStreams w Status.Rds")

# Create predictor Catchment Area variables that will maintain meaningful y-intercepts
xbar_CA<-mean(SimStreams_All$CA)
SimStreams_All$delta_CA<-SimStreams_All$CA - xbar_CA
xbar_Farming<-mean(SimStreams_All$Farming)
SimStreams_All$delta_Farming<-SimStreams_All$Farming - xbar_Farming

# Fit a model with common slope for CA and Farming and separate intercepts for each Ecoregion
EH0_Ecoregion_CA<-quap(
  alist(
    zNMDS2 ~ dnorm(mu,sigma),
    mu<-a_fit[Ecoregion] + b_CA*(delta_CA),
    a_fit[Ecoregion] ~ dnorm(50,10),
    b_CA ~ dnorm(1,1),
    sigma ~ dunif(0,25)
  ),
  data=SimStreams_All)
precis(EH0_Ecoregion_CA,depth=2,digits=4)

# Fit a model with common slope for CA and Farming and separate intercepts for each Ecoregion
EH1_Ecoregion_CA_Farming<-quap(
  alist(
    zNMDS2 ~ dnorm(mu,sigma),
    mu<-a_fit[Ecoregion] + b_CA*(delta_CA) + b_Farming*(delta_Farming),
    a_fit[Ecoregion] ~ dnorm(50,10),
    b_CA ~ dnorm(1,1),
    b_Farming ~ dnorm(1,1),
    sigma ~ dunif(0,25)
  ),
  data=SimStreams_All)
precis(EH1_Ecoregion_CA_Farming,depth=2,digits=4)

# Fit a model with different slope of Farming for each group -
# interaction between age group and weight
EH2_Interaction_Ecoregion<-quap(
  alist(
    zNMDS2 ~ dnorm(mu,sigma),
    mu<-a_fit[Ecoregion] + b_CA*(delta_CA) + b_Farming[Ecoregion]*(delta_Farming),
    a_fit[Ecoregion] ~ dnorm(50,10),
    b_CA ~ dnorm(1,1),
    b_Farming[Ecoregion] ~ dnorm(1,1),
    sigma ~ dunif(0,25)
  ),
  data=SimStreams_All)
precis(EH2_Interaction_Ecoregion,depth=3,digits=4)

# Now compare the models - smallest WAIC wins!
compare(EH0_Ecoregion_CA,EH1_Ecoregion_CA_Farming,EH2_Interaction_Ecoregion,func=WAIC)
plot(compare(EH0_Ecoregion_CA,EH1_Ecoregion_CA_Farming,EH2_Interaction_Ecoregion))
```

## SimStreams 4_4b Diagnosis Model NMDS2 from CA Ecoregion Farming.R—Finding the Best Model that Uses Both the Natural and Human Activity Environments to Explain Variation in NMDS2 Among Reference and Test Sites.

```
# Building a Bayesian model for ecosystem health (EH) in streams

# Load rethinking package
library(rethinking)

# Load data for modelling
SimStreams_All<-readRDS("./SimStreams/Datasets/SimStreams w Status.Rds")

# Create predictor Catchment Area variables that will maintain meaningful y-intercepts
xbar_CA<-mean(SimStreams_All$CA)
SimStreams_All$delta_CA<-SimStreams_All$CA - xbar_CA
xbar_Farming<-mean(SimStreams_All$Farming)
SimStreams_All$delta_Farming<-SimStreams_All$Farming - xbar_Farming

# Fit a model with common slope for CA and Farming and separate intercepts for each Ecoregion
EH0_Ecoregion_CA<-quap(
  alist(
    zNMDS2 ~ dnorm(mu,sigma),
    mu<-a_fit[Ecoregion] + b_CA*(delta_CA),
    a_fit[Ecoregion] ~ dnorm(50,10),
    b_CA ~ dnorm(1,1),
    sigma ~ dunif(0,25)
  ),
  data=SimStreams_All)
precis(EH0_Ecoregion_CA,depth=2,digits=4)

# Fit a model with common slope for CA and Farming and separate intercepts for each Ecoregion
EH1_Ecoregion_CA_Farming<-quap(
  alist(
    zNMDS2 ~ dnorm(mu,sigma),
    mu<-a_fit[Ecoregion] + b_CA*(delta_CA) + b_Farming*(delta_Farming),
    a_fit[Ecoregion] ~ dnorm(50,10),
    b_CA ~ dnorm(1,1),
    b_Farming ~ dnorm(1,1),
    sigma ~ dunif(0,25)
  ),
  data=SimStreams_All)
precis(EH1_Ecoregion_CA_Farming,depth=2,digits=4)

# Fit a model with different slope of Farming for each group -
# interaction between age group and weight
EH2_Interaction_Ecoregion<-quap(
  alist(
    zNMDS2 ~ dnorm(mu,sigma),
    mu<-a_fit[Ecoregion] + b_CA*(delta_CA) + b_Farming[Ecoregion]*(delta_Farming),
    a_fit[Ecoregion] ~ dnorm(50,10),
    b_CA ~ dnorm(1,1),
    b_Farming[Ecoregion] ~ dnorm(1,1),
    sigma ~ dunif(0,25)
  ),
  data=SimStreams_All)
precis(EH2_Interaction_Ecoregion,depth=3,digits=4)

# Now compare the models - smallest WAIC wins!
compare(EH0_Ecoregion_CA,EH1_Ecoregion_CA_Farming,EH2_Interaction_Ecoregion,func=WAIC)
plot(compare(EH0_Ecoregion_CA,EH1_Ecoregion_CA_Farming,EH2_Interaction_Ecoregion))
```

# *GLBEAST*

## GLBEAST 4_1a Bioassessment Taxon Richness.R—Comparing Actual Taxon Richness at Test and Reference Sites to the Prediction from the Reference Condition Model.

```
library(rethinking)
library(tidyverse)

# 1. Prepare data for fitting
GLBEAST<-readRDS("./GLBEAST/Datasets/GLBEAST RCA2_trim.Rds")
GLBEAST$log_Fetch<-log(1+GLBEAST$Fetch)
GLBEAST_Ref<-subset(GLBEAST,StatusRCA=="R")
GLBEAST_Test<-subset(GLBEAST,StatusRCA=="T")

# Create predictor Catchment Area variables that will maintain meaningful y-intercepts
xbar_log_Fetch<-mean(GLBEAST_Ref$log_Fetch)
GLBEAST_Ref$delta_log_Fetch<-GLBEAST_Ref$log_Fetch - xbar_log_Fetch
GLBEAST_Test$delta_log_Fetch<-GLBEAST_Test$log_Fetch - xbar_log_Fetch
xbar_Fetch<-mean(GLBEAST_Ref$Fetch)
GLBEAST_Ref$delta_Fetch<-GLBEAST_Ref$Fetch - xbar_Fetch
GLBEAST_Test$delta_Fetch<-GLBEAST_Test$Fetch - xbar_Fetch

# 2 Fit best model for Richness
EH3_GreatLake_log_Fetch<-quap(
  alist(
    Richness ~ dnorm(mu,sigma),
    mu<-a_fit[GreatLake] + b*(delta_log_Fetch),
    a_fit[GreatLake] ~ dnorm(20,10),
    b ~ dnorm(1,1),
    sigma ~ dunif(0,10)
  ),
  data=GLBEAST_Ref)
precis(EH3_GreatLake_log_Fetch,depth=2,digits=4)

# 2 Given the best guess at the parameters, figure out likelihood of actual values of
# Richness given GreatLake and Fetch
# Simulate 1000 Predicted Values for every Reference site
Richness_Pred<-link(EH3_GreatLake_log_Fetch,data=GLBEAST_Ref)
# Calculate Mean Predicted Value for every Reference site
GLBEAST_Ref$mu.Richness_Pred<-apply(Richness_Pred,2,mean)
GLBEAST_Ref$Richness_Resid<-GLBEAST_Ref$Richness - GLBEAST_Ref$mu.Richness_Pred
# Simulate 1000 Predicted Value for every Test site
Richness_Pred<-link(EH3_GreatLake_log_Fetch,data=GLBEAST_Test)
# Calculate Mean Predicted Value for every Test site
GLBEAST_Test$mu.Richness_Pred<-apply(Richness_Pred,2,mean)
GLBEAST_Test$Richness_Resid<-GLBEAST_Test$Richness - GLBEAST_Test$mu.Richness_Pred

GLBEAST_All_Bioass<-rbind(GLBEAST_Ref,GLBEAST_Test)

# 3 Build cdf of residual values and compare actual test and ref sites)
Resid_cdf<-ecdf(GLBEAST_Ref$Richness_Resid)
GLBEAST_All_Bioass$lpost<-Resid_cdf(GLBEAST_All_Bioass$Richness_Resid)
GLBEAST_All_Bioass$assess_post_Richness<-ifelse(GLBEAST_All_Bioass$lpost < 0.10,"VLOW",
       ifelse(GLBEAST_All_Bioass$lpost < 0.20, "LOW",
              ifelse(GLBEAST_All_Bioass$lpost > 0.90,"VHIGH",
                     ifelse(GLBEAST_All_Bioass$lpost > 0.80, "HIGH", "OK"))))
GLBEAST_All_Bioass$assess_post_Richness <- factor(GLBEAST_All_Bioass$assess_post_Richness, levels
= c("VLOW","LOW","OK","HIGH","VHIGH"))
GLBEAST_All_Bioass$StatusRCA<-factor(GLBEAST_All_Bioass$StatusRCA,
                                     levels = c("R","T"),
                                     labels = c("Reference", "Test"))
```

```
ggplot(GLBEAST_All_Bioass,aes(assess_post_Richness,group=StatusRCA,fill=StatusRCA)) +
  geom_bar(aes(y=..prop..),position="dodge") +
  theme(axis.text = element_text(color = "blue", size = 13)) +
  theme(axis.title = element_text(color = "blue", size = 14,face="bold"),
        strip.text = element_text(size=14, color = "blue")) +
  theme(
    legend.position = c(.42,.95),
    legend.justification = c(0, 1),
    legend.text = element_text(size = 13),
    legend.title = element_text(size = 13,face="bold")
  ) +
  scale_y_continuous(labels = scales::percent) +
  facet_wrap(~GreatLake) +
  xlab("Assessment") +
  ylab("% Sites")
scale_x_discrete(drop=FALSE)

write.csv(GLBEAST_All_Bioass,"./GLBEAST/Datasets/GLBEAST Bioassessment Richness.csv")
```

## GLBEAST 4_1b Bioassessment Community Composition.R—Comparing Actual Community Composition (Ordination Scores) at Test and Reference Sites to the Prediction from the Reference Condition Model.

```
library(rethinking)
library(tidyverse)

# 1. Prepare data for fitting
GLBEAST<-readRDS("./GLBEAST/Datasets/GLBEAST RCA2_trim.Rds")
GLBEAST$log_Fetch<-log(1+GLBEAST$Fetch)
GLBEAST_Ref<-subset(GLBEAST,StatusRCA=="R")
GLBEAST_Test<-subset(GLBEAST,StatusRCA=="T")

# Create predictor Catchment Area variables that will maintain meaningful y-intercepts
xbar_log_Fetch<-mean(GLBEAST_Ref$log_Fetch)
GLBEAST_Ref$delta_log_Fetch<-GLBEAST_Ref$log_Fetch - xbar_log_Fetch
GLBEAST_Test$delta_log_Fetch<-GLBEAST_Test$log_Fetch - xbar_log_Fetch
xbar_Fetch<-mean(GLBEAST_Ref$Fetch)
GLBEAST_Ref$delta_Fetch<-GLBEAST_Ref$Fetch - xbar_Fetch
GLBEAST_Test$delta_Fetch<-GLBEAST_Test$Fetch - xbar_Fetch


# 2 Fit a model for NMDS1
EH4_Interaction_NMDS1<-quap(
  alist(
    z_PA_MDS1 ~ dnorm(mu,sigma),
    mu<-a_fit[GreatLake] + b[GreatLake]*(delta_log_Fetch),
    a_fit[GreatLake] ~ dnorm(60,20),
    b[GreatLake] ~ dnorm(1,1),
    sigma ~ dunif(0,10)
  ),
  data=GLBEAST_Ref)
precis(EH4_Interaction_NMDS1,depth=3,digits=4)

# 3 Given the best guess at the parameters, figure out likelihood
# of actual values of Spp given GreatLake and Catchment Area
# Simulate 1000 Predicted Values for every Reference site
z_PA_MDS1_Pred<-link(EH4_Interaction_NMDS1,data=GLBEAST_Ref)
# Calculate Mean Predicted Value for every Reference site
GLBEAST_Ref$mu.z_PA_MDS1_Pred<-apply(z_PA_MDS1_Pred,2,mean)
GLBEAST_Ref$z_PA_MDS1_Resid<-GLBEAST_Ref$z_PA_MDS1 - GLBEAST_Ref$mu.z_PA_MDS1_Pred

# Simulate 1000 Predicted Value for every Test site
z_PA_MDS1_Pred<-link(EH4_Interaction_NMDS1,data=GLBEAST_Test)
# Calculate Mean Predicted Value for every Test site
GLBEAST_Test$mu.z_PA_MDS1_Pred<-apply(z_PA_MDS1_Pred,2,mean)
GLBEAST_Test$z_PA_MDS1_Resid<-GLBEAST_Test$z_PA_MDS1 - GLBEAST_Test$mu.z_PA_MDS1_Pred

GLBEAST_All_Bioass<-rbind(GLBEAST_Ref,GLBEAST_Test)

# 3 Build cdf of residual values and compare actual test and ref sites)
Resid_cdf_NMDS1<-ecdf(GLBEAST_Ref$z_PA_MDS1_Resid)
GLBEAST_All_Bioass$lpost_NMDS1<-Resid_cdf_NMDS1(GLBEAST_All_Bioass$z_PA_MDS1_Resid)
GLBEAST_All_Bioass$assess_post_NMDS1<-ifelse(GLBEAST_All_Bioass$lpost_NMDS1 < 0.10,"VLOW",
                                     ifelse(GLBEAST_All_Bioass$lpost_NMDS1 < 0.20, "LOW",
                                            ifelse(GLBEAST_All_Bioass$lpost_NMDS1 >
0.90,"VHIGH",
                                                   ifelse(GLBEAST_All_Bioass$lpost_NMDS1 > 0.80,
"HIGH", "OK"))))
GLBEAST_All_Bioass$assess_post_NMDS1 <- factor(GLBEAST_All_Bioass$assess_post_NMDS1, levels =
c("VLOW","LOW","OK","HIGH","VHIGH"))
```

```
GLBEAST_Ref<-subset(GLBEAST_All_Bioass,StatusRCA=="R")
GLBEAST_Test<-subset(GLBEAST_All_Bioass,StatusRCA=="T")

# 4 Fit a model for NMDS2
EH3_GreatLake_log_Fetch_NMDS2<-quap(
  alist(
    z_PA_MDS2 ~ dnorm(mu,sigma),
    mu<-a_fit[GreatLake] + b*(delta_log_Fetch),
    a_fit[GreatLake] ~ dnorm(20,10),
    b ~ dnorm(1,1),
    sigma ~ dunif(0,10)
  ),
  data=GLBEAST_Ref)
precis(EH3_GreatLake_log_Fetch_NMDS2,depth=3,digits=4)

# 5 Given the best guess at the parameters, figure out likelihood of actual values of
# Spp given GreatLake and Catchment Area
# Simulate 1000 Predicted Values for every Reference site
z_PA_MDS2_Pred<-link(EH3_GreatLake_log_Fetch_NMDS2,data=GLBEAST_Ref)
# Calculate Mean Predicted Value for every Reference site
GLBEAST_Ref$mu.z_PA_MDS2_Pred<-apply(z_PA_MDS2_Pred,2,mean)
GLBEAST_Ref$z_PA_MDS2_Resid<-GLBEAST_Ref$z_PA_MDS2 - GLBEAST_Ref$mu.z_PA_MDS2_Pred

# Simulate 1000 Predicted Value for every Test site
z_PA_MDS2_Pred<-link(EH3_GreatLake_log_Fetch_NMDS2,data=GLBEAST_Test)
# Calculate Mean Predicted Value for every Test site
GLBEAST_Test$mu.z_PA_MDS2_Pred<-apply(z_PA_MDS2_Pred,2,mean)
GLBEAST_Test$z_PA_MDS2_Resid<-GLBEAST_Test$z_PA_MDS2 - GLBEAST_Test$mu.z_PA_MDS2_Pred

GLBEAST_All_Bioass<-rbind(GLBEAST_Ref,GLBEAST_Test)

# 6 Build cdf of residual values and compare actual test and ref sites)
Resid_cdf_NMDS2<-ecdf(GLBEAST_Ref$z_PA_MDS2_Resid)
GLBEAST_All_Bioass$lpost_NMDS2<-Resid_cdf_NMDS2(GLBEAST_All_Bioass$z_PA_MDS2_Resid)
GLBEAST_All_Bioass$assess_post_NMDS2<-ifelse(GLBEAST_All_Bioass$lpost_NMDS2 < 0.10,"VLOW",
                                            ifelse(GLBEAST_All_Bioass$lpost_NMDS2 < 0.20, "LOW",
                                                   ifelse(GLBEAST_All_Bioass$lpost_NMDS2 >
0.90,"VHIGH",
                                                          ifelse(GLBEAST_All_Bioass$lpost_NMDS2 > 0.80,
"HIGH", "OK"))))
GLBEAST_All_Bioass$assess_post_NMDS2 <- factor(GLBEAST_All_Bioass$assess_post_NMDS2, levels =
c("VLOW","LOW","OK","HIGH","VHIGH"))


# 7 Do combined assessment of NMDS1 NMDS2
GLBEAST_All_Bioass$assess_post_Community<-ifelse((GLBEAST_All_Bioass$assess_post_NMDS1 == "VLOW"
| GLBEAST_All_Bioass$assess_post_NMDS2 == "VLOW"),"VLOW",
                                                ifelse((GLBEAST_All_Bioass$assess_post_NMDS1 == "LOW" &
GLBEAST_All_Bioass$assess_post_NMDS2 == "LOW"),"VLOW",
                                                       ifelse((GLBEAST_All_Bioass$assess_post_NMDS1 ==
"LOW" | GLBEAST_All_Bioass$assess_post_NMDS2 == "LOW"),"LOW",

ifelse((GLBEAST_All_Bioass$assess_post_NMDS1 == "OK" & GLBEAST_All_Bioass$assess_post_NMDS2 ==
"OK"),"OK",

ifelse((GLBEAST_All_Bioass$assess_post_NMDS1 == "HIGH" & GLBEAST_All_Bioass$assess_post_NMDS2 ==
"HIGH"),"VHIGH",

ifelse((GLBEAST_All_Bioass$assess_post_NMDS1 == "HIGH" | GLBEAST_All_Bioass$assess_post_NMDS2 ==
"HIGH"),"HIGH",

ifelse((GLBEAST_All_Bioass$assess_post_NMDS1 == "VHIGH" | GLBEAST_All_Bioass$assess_post_NMDS2 ==
"VHIGH"),"VHIGH","ND")))))))
GLBEAST_All_Bioass$assess_post_Community <- factor(GLBEAST_All_Bioass$assess_post_Community,
levels = c("VLOW","LOW","OK","HIGH","VHIGH"))

GLBEAST_All_Bioass$StatusRCA<-factor(GLBEAST_All_Bioass$StatusRCA,
                                     levels = c("R","T"),
                                     labels = c("Reference", "Test"))
```

```
ggplot(GLBEAST_All_Bioass,aes(assess_post_Community,group=StatusRCA,fill=StatusRCA)) +
  geom_bar(aes(y=..prop..),position="dodge") +
  theme(axis.text = element_text(color = "blue", size = 13)) +
  theme(axis.title = element_text(color = "blue", size = 14,face="bold"),
        strip.text = element_text(size=14, color = "blue")) +
  theme(
    legend.position = c(.42,.45),
    legend.justification = c(0, 1),
    legend.text = element_text(size = 13),
    legend.title = element_text(size = 13,face="bold")
  ) +
  scale_y_continuous(labels = scales::percent) +
  facet_wrap(~GreatLake) +
  xlab("Assessment") +
  ylab("% Sites")
scale_x_discrete(drop=FALSE)

write.csv(GLBEAST_All_Bioass,"./GLBEAST/Datasets/GLBEAST Bioassessment NMDS.csv")
```

## GLBEAST 4_2 Diagnosis Model Richness HAG.R—Finding the Best Model that Uses Both the Natural and Human Activity Environments to Explain Variation in Taxon Richness Among Reference and Test Sites.

```
library(rethinking)
library(tidyverse)
# From McElreath 2020 pp78-91

# Prepare data for fitting
GLBEAST<-readRDS("./GLBEAST/Datasets/GLBEAST RCA2_trim.Rds")
GLBEAST$log_Fetch<-log(1+GLBEAST$Fetch)
# Create predictor Catchment Area variables that will maintain meaningful y-intercepts
xbar_log_Fetch<-mean(GLBEAST$log_Fetch)
GLBEAST$delta_log_Fetch<-GLBEAST$log_Fetch - xbar_log_Fetch
xbar_HAG_PC1<-mean(GLBEAST$HAG_PC1)
GLBEAST$delta_HAG_PC1<-GLBEAST$HAG_PC1 - xbar_HAG_PC1

# Fit a model with common slope for Fetch and separate intercepts for each lake
EH0_GreatLake_logEF<-quap(
  alist(
    Richness ~ dnorm(mu,sigma),
    mu<-a_fit[GreatLake] + b*(delta_log_Fetch),
    a_fit[GreatLake] ~ dnorm(20,10),
    b ~ dnorm(1,1),
    sigma ~ dunif(0,10)
  ),
  data=GLBEAST)
precis(EH0_GreatLake_logEF,depth=3,digits=4)

# Fit a model with common slope for Fetch and HAG_PC1
EH1_GreatLake_logEF_HAG<-quap(
  alist(
    Richness ~ dnorm(mu,sigma),
    mu<-a_fit[GreatLake] + b_EF*(delta_log_Fetch) + b_HAG*(delta_HAG_PC1),
    a_fit[GreatLake] ~ dnorm(20,10),
    b_EF ~ dnorm(1,1),
    b_HAG ~ dnorm(1,1),
    sigma ~ dunif(0,10)
  ),
  data=GLBEAST)
precis(EH1_GreatLake_logEF_HAG,depth=3,digits=4)

# Fit a model with common slope for Fetch
# and separate slopes for HAG_PC1
EH2_Interaction_GreatLake_HAG<-quap(
  alist(
    Richness ~ dnorm(mu,sigma),
    mu<-a_fit[GreatLake] + b_EF*(delta_log_Fetch) + b_HAG[GreatLake]*(delta_HAG_PC1),
    a_fit[GreatLake] ~ dnorm(20,10),
    b_EF ~ dnorm(1,1),
    b_HAG[GreatLake] ~ dnorm(1,1),
    sigma ~ dunif(0,10)
  ),
  data=GLBEAST)
precis(EH2_Interaction_GreatLake_HAG,depth=3,digits=4)

# Now compare the models - smallest WAIC wins!
compare(EH0_GreatLake_logEF,EH1_GreatLake_logEF_HAG,EH2_Interaction_GreatLake_HAG,func=WAIC)
plot(compare(EH0_GreatLake_logEF,EH1_GreatLake_logEF_HAG,EH2_Interaction_GreatLake_HAG))
```

## GLBEAST 4_3 Diagnosis Best Model PredPlot HAG Taxon Richness.R—Showing Model Fit of Variation in Taxon Richness with the Natural and Human Activity Environments.

```
library(rethinking)
library(tidyverse)
# From McElreath 2020 pp78-91

# Load data for modelling
GLBEAST<-readRDS("./GLBEAST/Datasets/GLBEAST RCA2_trim.Rds")
GLBEAST$log_Fetch<-log(1+GLBEAST$Fetch)
# Create predictor Catchment Area variables that will maintain meaningful y-intercepts
xbar_log_Fetch<-mean(GLBEAST$log_Fetch)
GLBEAST$delta_log_Fetch<-GLBEAST$log_Fetch - xbar_log_Fetch
xbar_HAG<-mean(GLBEAST$HAG_PC1)
GLBEAST$delta_HAG<-GLBEAST$HAG_PC1 - xbar_HAG

# 2. Fit best model
EH2_Interaction_GreatLake_HAG<-quap(
  alist(
    Richness ~ dnorm(mu,sigma),
    mu<-a_fit[GreatLake] + b_EF*(delta_log_Fetch) + b_HAG[GreatLake]*(delta_HAG),
    a_fit[GreatLake] ~ dnorm(20,10),
    b_EF ~ dnorm(1,1),
    b_HAG[GreatLake] ~ dnorm(1,1),
    sigma ~ dunif(0,10)
  ),
  data=GLBEAST)
precis(EH2_Interaction_GreatLake_HAG,depth=3,digits=4)

# 3 Given the best guess at the parameters, figure out likelihood of actual values of
# Richness given GreatLake, Effective Fetch, and Human Activity Gradient

# Simulate 1000 Predicted Values for every site
Richness_Pred<-link(EH2_Interaction_GreatLake_HAG,data=GLBEAST)
# Calculate Mean Predicted Value for every Reference site
GLBEAST$mu.Richness_Pred<-apply(Richness_Pred,2,mean)
GLBEAST$Richness_Resid<-GLBEAST$Richness - GLBEAST$mu.Richness_Pred

# Create a dataset [SimStreams] grid of Catchment Area (100 to 1000ha) and Ecoregion (1,2) values
Fetch.list<-seq(from=0,to=100,length.out=25)
GreatLake.list<-c(1,2,3,4)
HAG.list<-seq(from=-5,to=5,length.out=25)
GLBEAST_Pred<-expand.grid(GreatLake=GreatLake.list,Fetch=Fetch.list,HAG=HAG.list)
GLBEAST_Pred$delta_HAG<-GLBEAST_Pred$HAG - xbar_HAG
GLBEAST_Pred$log_Fetch<-log(1+GLBEAST_Pred$Fetch)
GLBEAST_Pred$delta_log_Fetch<-GLBEAST_Pred$log_Fetch - xbar_log_Fetch
GLBEAST_Sim<-sim(EH2_Interaction_GreatLake_HAG,data=GLBEAST_Pred,n=100)
GLBEAST_Pred$Richness_Pred<-apply(GLBEAST_Sim,2,mean)

# First plot fitted model for Fetch
ggplot(GLBEAST_Pred,aes(Fetch,Richness_Pred,color=factor(GreatLake)))+
  theme(axis.text = element_text(color = "blue", size = 12)) +
  theme(axis.title = element_text(color = "blue", size = 14,face="bold")) +
  xlab("Effective Fetch") +
  ylab("Taxon Richness") +
  geom_smooth(linetype="dashed",show.legend=FALSE) +
  scale_color_manual(values = c( "blue", "yellow", "black","red"))
```

```
# Then plot fitted model for Farming
ggplot(GLBEAST_Pred,aes(HAG,Richness_Pred,color=factor(GreatLake)))+
  theme(axis.text = element_text(color = "blue", size = 12)) +
  theme(axis.title = element_text(color = "blue", size = 14,face="bold")) +
  xlab("HAG_PC1") +
  ylab("Species Richness") +
  geom_smooth(linetype="dashed",show.legend=FALSE) +
  scale_color_manual(values = c( "blue", "yellow", "black","red"))

# Then plot Farming v Catchment Area
ggplot(GLBEAST,aes(Fetch,HAG_PC1,color=factor(GreatLake)))+
  theme(axis.text = element_text(color = "blue", size = 12)) +
  theme(axis.title = element_text(color = "blue", size = 14,face="bold")) +
  xlab("Fetch") +
  ylab("HAG") +
  geom_point(show.legend=FALSE,size=-1) +
  scale_color_manual(values = c("black","black","red","red"))+
  geom_text(data=GLBEAST,aes(color=factor(GreatLake)),
          x=GLBEAST$Fetch,y=GLBEAST$HAG_PC1,fontface="bold",
          show.legend=FALSE,label=GLBEAST$Richness,size=3)
```

## GLBEAST 4_4a Rethinking Model NMDS1 HAG.R—Finding the Best Model that Uses Both the Natural and Human Activity Environments to Explain Variation in NMDS1 Among Reference and Test Sites.

```
library(rethinking)
library(tidyverse)
# From McElreath 2020 pp78-91

# Prepare data for fitting
GLBEAST<-readRDS("./GLBEAST/Datasets/GLBEAST RCA2_trim.Rds")
GLBEAST$log_Fetch<-log(1+GLBEAST$Fetch)
# Create predictor Catchment Area variables that will maintain meaningful y-intercepts
xbar_log_Fetch<-mean(GLBEAST$log_Fetch)
GLBEAST$delta_log_Fetch<-GLBEAST$log_Fetch - xbar_log_Fetch
xbar_HAG_PC1<-mean(GLBEAST$HAG_PC1)
GLBEAST$delta_HAG_PC1<-GLBEAST$HAG_PC1 - xbar_HAG_PC1

# Fit a model with common slope for Fetch and separate intercepts for each lake
EH0_GreatLake_logEF<-quap(
  alist(
    z_PA_MDS1 ~ dnorm(mu,sigma),
    mu<-a_fit[GreatLake] + b*(delta_log_Fetch),
    a_fit[GreatLake] ~ dnorm(20,10),
    b ~ dnorm(1,1),
    sigma ~ dunif(0,10)
  ),
  data=GLBEAST)
precis(EH0_GreatLake_logEF,depth=3,digits=4)

# Fit a model with common slope for Fetch and HAG_PC1
EH1_GreatLake_logEF_HAG<-quap(
  alist(
    z_PA_MDS1 ~ dnorm(mu,sigma),
    mu<-a_fit[GreatLake] + b_EF*(delta_log_Fetch) + b_HAG*(delta_HAG_PC1),
    a_fit[GreatLake] ~ dnorm(20,10),
    b_EF ~ dnorm(1,1),
    b_HAG ~ dnorm(1,1),
    sigma ~ dunif(0,10)
  ),
  data=GLBEAST)
precis(EH1_GreatLake_logEF_HAG,depth=3,digits=4)

# Fit a model with common slope for Fetch
# and separate slopes for HAG_PC1
EH2_Interaction_GreatLake_HAG<-quap(
  alist(
    z_PA_MDS1 ~ dnorm(mu,sigma),
    mu<-a_fit[GreatLake] + b_EF*(delta_log_Fetch) + b_HAG[GreatLake]*(delta_HAG_PC1),
    a_fit[GreatLake] ~ dnorm(20,10),
    b_EF ~ dnorm(1,1),
    b_HAG[GreatLake] ~ dnorm(1,1),
    sigma ~ dunif(0,10)
  ),
  data=GLBEAST)
precis(EH2_Interaction_GreatLake_HAG,depth=3,digits=4)

# Now compare the models - smallest WAIC wins!
compare(EH0_GreatLake_logEF,EH1_GreatLake_logEF_HAG,EH2_Interaction_GreatLake_HAG,func=WAIC)
plot(compare(EH0_GreatLake_logEF,EH1_GreatLake_logEF_HAG,EH2_Interaction_GreatLake_HAG))
```

## GLBEAST 4_4b Rethinking Model NMDS2 HAG.R—Finding the Best Model that Uses Both the Natural and Human Activity Environments to Explain Variation in NMDS2 Among Reference and Test Sites.

```
library(rethinking)
library(tidyverse)
# From McElreath 2020 pp78-91

# Prepare data for fitting
GLBEAST<-readRDS("./GLBEAST/Datasets/GLBEAST RCA2_trim.Rds")
GLBEAST$log_Fetch<-log(1+GLBEAST$Fetch)
# Create predictor Catchment Area variables that will maintain meaningful y-intercepts
xbar_log_Fetch<-mean(GLBEAST$log_Fetch)
GLBEAST$delta_log_Fetch<-GLBEAST$log_Fetch - xbar_log_Fetch
xbar_HAG_PC1<-mean(GLBEAST$HAG_PC1)
GLBEAST$delta_HAG_PC1<-GLBEAST$HAG_PC1 - xbar_HAG_PC1

# Fit a model with common slope for Fetch and separate intercepts for each lake
EH0_GreatLake_logEF<-quap(
  alist(
    z_PA_MDS2 ~ dnorm(mu,sigma),
    mu<-a_fit[GreatLake] + b*(delta_log_Fetch),
    a_fit[GreatLake] ~ dnorm(20,10),
    b ~ dnorm(1,1),
    sigma ~ dunif(0,10)
  ),
  data=GLBEAST)
precis(EH0_GreatLake_logEF,depth=3,digits=4)

# Fit a model with common slope for Fetch and HAG_PC1
EH1_GreatLake_logEF_HAG<-quap(
  alist(
    z_PA_MDS2 ~ dnorm(mu,sigma),
    mu<-a_fit[GreatLake] + b_EF*(delta_log_Fetch) + b_HAG*(delta_HAG_PC1),
    a_fit[GreatLake] ~ dnorm(20,10),
    b_EF ~ dnorm(1,1),
    b_HAG ~ dnorm(1,1),
    sigma ~ dunif(0,10)
  ),
  data=GLBEAST)
precis(EH1_GreatLake_logEF_HAG,depth=3,digits=4)

# Fit a model with common slope for Fetch
# and separate slopes for HAG_PC1
EH2_Interaction_GreatLake_HAG<-quap(
  alist(
    z_PA_MDS2 ~ dnorm(mu,sigma),
    mu<-a_fit[GreatLake] + b_EF*(delta_log_Fetch) + b_HAG[GreatLake]*(delta_HAG_PC1),
    a_fit[GreatLake] ~ dnorm(20,10),
    b_EF ~ dnorm(1,1),
    b_HAG[GreatLake] ~ dnorm(1,1),
    sigma ~ dunif(0,10)
  ),
  data=GLBEAST)
precis(EH2_Interaction_GreatLake_HAG,depth=3,digits=4)

# Now compare the models - smallest WAIC wins!
compare(EH0_GreatLake_logEF,EH1_GreatLake_logEF_HAG,EH2_Interaction_GreatLake_HAG,func=WAIC)
plot(compare(EH0_GreatLake_logEF,EH1_GreatLake_logEF_HAG,EH2_Interaction_GreatLake_HAG))
```

## References

Kilgour, B.W., Somers, K.M., Barrett, T.J., Munkittrick, K.R., Francis, A.P.: Testing against “normal” with environmental data. Integr. Environ. Assess. Manag. **13**, 188–197 (2017)

McElreath, R.: Statistical Rethinking: A Bayesian course with examples in R and STAN. 2nd edition. Chapman & Hall/CRC (2020)

# Chapter 5
# Modeling Temporal Change in Freshwater Ecosystems

As in all ecosystems, the natural and human activity freshwater environments are constantly changing. Most lakes in temperate regions develop a classic depth gradient in temperature by late summer (Fig. 5.1), and flow in a stream can vary by orders of magnitude within a 24 h period (Fig. 5.2). The bioassessment examples I've presented in this book have been typical of many programs … the data are treated as if the biota and its environmental drivers vary from site to site but are pretty much static over time. But our methods of evaluating ecosystem health should not be limited to detecting degradation (or rehabilitation) over short time periods. Many programs need to be able to discern the effects of important temporal stressors like climate change over annual, decadal, and longer time periods.

A few years ago, Trefor Reynoldson and I were asked to review decades of bioassessment data from hundreds of stream sites in British Columbia, Canada, for evidence of climate change. We both felt it was an intriguing project and since we (and especially Trefor) had been part of the bioassessment program in BC from its start in the early 90 s, we fully expected to spend most of our time on the project coming up with clever ways to describe *how* climate change had affected BC streams. We ended up re-learning some hard lessons and being disappointed. I think Trefor would agree with my summary of our work…"it's complicated" (Bailey and Reynoldson 2023).

The first, difficult experience we had that will be familiar to many readers happens whenever some sort of re-analysis or synthesis is done of relatively big datasets, particularly bioassessment programs. The dirty secret is that the vast majority of effort is spent getting the data in shape rather than on novel, sophisticated analyses. I'm pretty sure this is not unique to the datasets I've worked with … bringing together environmental and biota data from a site, especially with an explicit time stamp, can be unusually hard. This isn't helped by the common practice of giving sites that have been sampled in different years unique site codes (e.g., "LakeBailey-1998-01"). This sounds like a trivial fix in Excel or Access, but multiply that across hundreds or thousands of records and it gets to be quite a pain, with hard to catch errors that can have major effects on the results. So, boring lesson one is: ***make sure***

 
R. C. Bailey, *Rethinking Bioassessment of Freshwater Ecosystem Health*,
https://doi.org/10.1007/978-3-032-12427-2_5

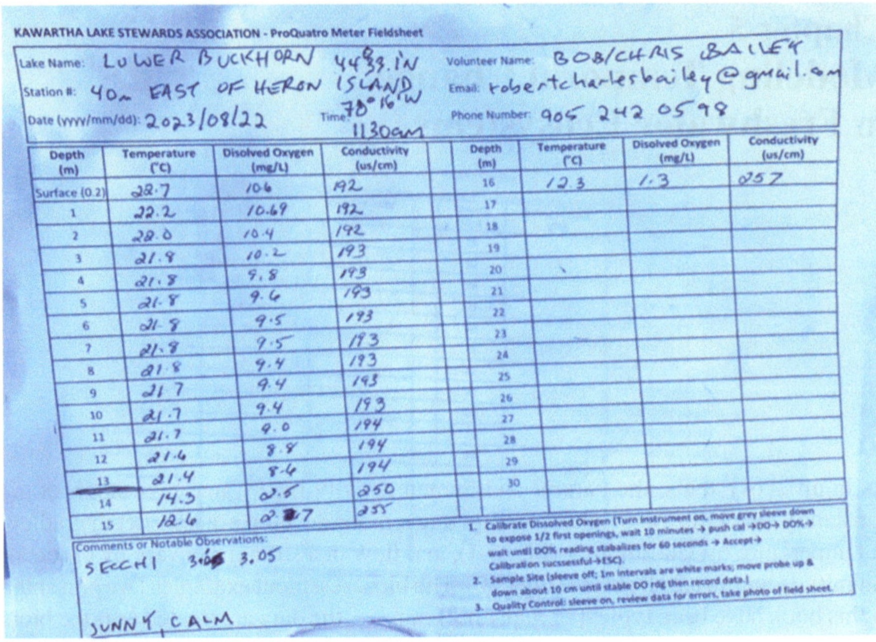

KAWARTHA LAKE STEWARDS ASSOCIATION - ProQuatro Meter Fieldsheet

Lake Name: LOWER BUCKHORN 44°33.1'N
Station #: 40m EAST OF HERON ISLAND 78°16'W
Date (yyyy/mm/dd): 2023/08/22 Time: 1130am
Volunteer Name: BOB/CHRIS BAILEY
Email: robertcharlesbailey@gmail.com
Phone Number: 905 242 0598

| Depth (m) | Temperature (°C) | Disolved Oxygen (mg/L) | Conductivity (us/cm) |
|---|---|---|---|
| Surface (0.2) | 22.7 | 10.6 | 192 |
| 1 | 22.2 | 10.69 | 192 |
| 2 | 22.0 | 10.4 | 192 |
| 3 | 21.8 | 10.2 | 193 |
| 4 | 21.8 | 9.8 | 193 |
| 5 | 21.8 | 9.6 | 193 |
| 6 | 21.8 | 9.5 | 193 |
| 7 | 21.8 | 9.5 | 193 |
| 8 | 21.8 | 9.4 | 193 |
| 9 | 21.7 | 9.4 | 193 |
| 10 | 21.7 | 9.4 | 193 |
| 11 | 21.7 | 9.0 | 194 |
| 12 | 21.6 | 8.8 | 194 |
| 13 | 21.4 | 8.4 | 194 |
| 14 | 14.3 | 2.5 | 250 |
| 15 | 12.6 | 2.7 | 255 |

| Depth (m) | Temperature (°C) | Disolved Oxygen (mg/L) | Conductivity (us/cm) |
|---|---|---|---|
| 16 | 12.3 | 1.3 | 257 |
| 17 | | | |
| 18 | | | |
| 19 | | | |
| 20 | | | |
| 21 | | | |
| 22 | | | |
| 23 | | | |
| 24 | | | |
| 25 | | | |
| 26 | | | |
| 27 | | | |
| 28 | | | |
| 29 | | | |
| 30 | | | |

Comments or Notable Observations:
SECCHI 3.05
SUNNY, CALM

1. Calibrate Dissolved Oxygen (Turn instrument on, move grey sleeve down to expose 1/2 first openings, wait 10 minutes → push cal → DO → DO% → wait until DO% reading stabalizes for 60 seconds → Accept → Calibration sucsessful → ESC).
2. Sample Site (sleeve off; 1m intervals are white marks; move probe up & down about 10 cm until stable DO rdg then record data.)
3. Quality Control: sleeve on, review data for errors, take photo of field sheet.

**Fig. 5.1** Vertical profile of temperature, dissolved oxygen, and conductivity in Lower Buckhorn Lake, Ontario, on August 22, 2023

**Fig. 5.2** Flow in a small stream draining urban storm sewers just after a heavy rain (left) and a few hours later (right)

***that bioassessment databases are structured for a variety of spatial and temporal analyses and aggregation.***

The other lesson we re-learned is one that Trefor and I have taught others many, many times over the years. ***Any study or program, including a bioassessment program, has to be designed to answer specific questions that are defined at the outset.*** This is nothing to do with avoiding "cheating" when doing statistical hypothesis tests…it's more basic than that. The bioassessment program in BC was well designed to evaluate Test sites to see if there was evidence that human activity had changed the biota in the ecosystem. It was *not* designed to assess temporal trends at individual Reference and Test sites, so we found a very small proportion of the hundreds of sites we looked at had been sampled in more than *three years*. It's pretty hard to find trends over time consistent with climate change when that's the nature of the dataset!

There were other lessons learned and re-learned in that study, and we did what we could to work around the limitations, but I'd like to suggest a few ideas for we can incorporate time into a bioassessment program either explicitly in the design or retrospectively, as Trefor and I did. I will look at time using the GLBEAST dataset, as it was collected over a 15 year period (2000–2015).

Adding time to the GLBEAST conceptual model (Fig. 5.3) illustrates one simple approach to dealing with time. "Year of sampling" may cause variation in the biota observed either because of temporal change in the natural environment or mediated by temporal changes in both human activity and its effects. In the case of GLBEAST, neither of the natural environment descriptors (Great Lake, Effective Fetch) are affected by time, but other descriptors could be. There have been major, year-to-year, ups and downs in water level in the Great Lakes, for example, which almost certainly affect the benthic invertebrate communities we see in nearshore ecosystems. And even the simple descriptors of human activity I have used with GLBEAST, roads, buildings, and people, changed substantially in some areas over the period of sampling. So we can add time (year of sampling) to both our Reference Condition and diagnosis models to see if there is any evidence of natural or human-influenced changes through time in the benthic community. We always have to remember, though, that just like the BC streams dataset that Trefor and I looked at, the GLBEAST programmed was not designed to detect temporal changes.

When Year is added to the Reference Condition model for taxon richness fitted in Chapter 3, the best model (Table 5.1) shows a similar decline (0.4 taxa/year) in richness at Reference sites in all four Great Lakes (Table 5.2).

The best diagnosis model is what I call Great_Lake_BACI…all the same predictors as in the human activity model from Chapter Four, but an interaction between human activity and year that is specific to each Great Lake.

Looking at the likely values of parameters in the BACI model (Table 5.4) enables an interpretation of how human activity has different consequences over time on the taxon richness in each Great Lake. In Lake Superior, there is evidence that increasing human activity increases taxon richness, and this effect stays consistent through time. I conclude this because in Table 5.4, the range of most likely values for the $\boldsymbol{\beta\_HA_{Superior}}$ parameter is well above zero (more human activity causes higher

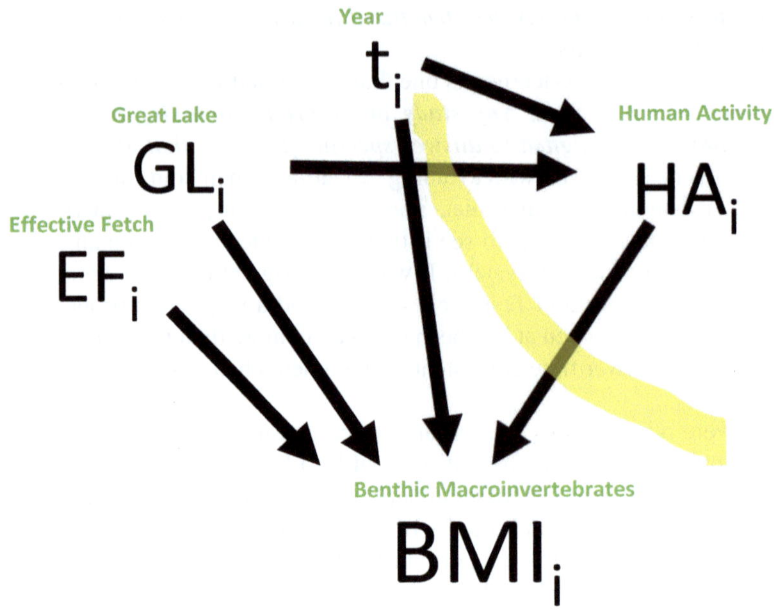


**Fig. 5.3** Conceptual model with temporal variability for benthic invertebrate communities in GLBEAST nearshore Great Lakes ecosystems

**Table 5.1** GLBEAST Taxon Richness: Choosing among Reference Condition models with year as an added cause. Best model is same effect of year in each of the four lakes (GreatLake_logEF_Year)

| | WAIC | SE | dWAIC | dSE | pWAIC | Weight |
|---|---|---|---|---|---|---|
| GreatLake_logEF_Year | 1593.1 | 37.21 | 0.0 | NA | 10.1 | 0.51 |
| GreatLake_Year Interaction | 1593.2 | 40.10 | 0.1 | 9.91 | 14.1 | 0.49 |
| GreatLake_logEF | 1628.8 | 36.42 | 35.7 | 12.67 | 9.2 | 0.00 |

**Table 5.2** Bayesian fit of Reference Condition model of variation in taxon richness among GLBEAST Reference sites with year as an added parameter shows a decline in taxon richness of 0.5 taxa/year in all four Great Lakes

| | Mean | Standard deviation | 5.5% | 94.5% |
|---|---|---|---|---|
| $\alpha_{Superior}$ | 8.40 | 0.46 | 7.66 | 9.13 |
| $\alpha_{Huron}$ | 7.77 | 0.55 | 6.90 | 8.65 |
| $\alpha_{Erie}$ | 18.44 | 1.19 | 16.53 | 20.35 |
| $\alpha_{Ontario}$ | 7.09 | 0.69 | 5.99 | 8.19 |
| $\beta_{fetch}$ | −1.01 | 0.27 | −1.45 | −0.57 |
| $\beta_{year}$ | −0.40 | 0.06 | −0.51 | −0.30 |
| $\sigma$ | 4.42 | 0.19 | 4.12 | 5.07 |

**Table 5.3** GLBEAST Taxon Richness: Choosing among human activity models with year as an added cause. Best model is same effect of year in each of the four lakes (GreatLake_Year Independent)

| | WAIC | SE | dWAIC | dSE | pWAIC | Weight |
|---|---|---|---|---|---|---|
| GreatLake_BACI | 3223.4 | 47.48 | 0.0 | NA | 16.2 | 0.62 |
| GreatLake_Year Independent | 3224.4 | 47.44 | 1.0 | 5.81 | 13.1 | 0.38 |
| GreatLake_NoYear | 3291.0 | 45.74 | 67.6 | 17.86 | 12.5 | 0.00 |

**Table 5.4** Bayesian fit of human activity model of variation in taxon richness among GLBEAST Reference and Test sites with year as an added parameter shows a decline in taxon richness of 0.5 taxa/year in all four Great Lakes

| | Mean | Standard deviation | 5.5% | 94.5% |
|---|---|---|---|---|
| $\alpha_{Superior}$ | 10.7645 | 0.3860 | 10.1476 | 11.3814 |
| $\alpha_{Huron}$ | 9.8536 | 0.6414 | 8.8284 | 10.8787 |
| $\alpha_{Erie}$ | 16.7731 | 0.9573 | 15.2432 | 18.3030 |
| $\alpha_{Ontario}$ | 8.3827 | 0.7177 | 7.2357 | 9.5298 |
| $\beta\_EF$ | −1.1479 | 0.2442 | −1.5381 | −0.7576 |
| $\beta\_HA_{Superior}$ | 0.5877 | 0.1242 | 0.3892 | 0.7863 |
| $\beta\_HA_{Huron}$ | 0.7519 | 0.2647 | 0.3289 | 1.1750 |
| $\beta\_HA_{Erie}$ | −1.2647 | 0.4036 | −1.9098 | −0.6196 |
| $\beta\_HA_{Ontario}$ | −0.5071 | 0.1848 | −0.8024 | −0.2117 |
| $\beta\_Year$ | −0.4967 | 0.0606 | −0.5936 | −0.3998 |
| $\beta_{BACI\_Superior}$ | −0.0437 | 0.0275 | −0.0876 | 0.0003 |
| $\beta_{BACI\_Huron}$ | −0.0921 | 0.0475 | −0.1680 | −0.0163 |
| $\beta_{BACI\_Erie}$ | −0.0109 | 0.0822 | −0.1423 | 0.1205 |
| $\beta_{BACI\_Ontario}$ | 0.0450 | 0.0196 | 0.0137 | 0.0763 |
| $\sigma$ | 4.6737 | 0.1426 | 4.4458 | 4.9017 |

taxon richness), and the range of likely values for the $\boldsymbol{\beta}_{BACI\_Superior}$ parameter include zero (time does not affect the effect of human activity on taxon richness). The same is true of Lake Erie, but it shows a reduction in taxon richness with human activity that is not changing over time. Taxon richness increase in Lake Huron and decrease in Lake Ontario with greater human activity are each modestly dampened through time. Such detailed interpretations of course are at the mercy of the amount of sampling in time and space in all four Great Lakes. Regardless of the model-fitting technique, nothing replaces a comprehensive sampling through time at a variety of natural and human activity environments in all four Great Lakes.

Just to be clear, I don't think my modeling approach here is the "silver bullet" when it comes to dealing with time in bioassessment studies. Conceptual models that include time, and conceptual models for bioassessment programs in general, get a lot more complicated than the simple examples I've shared here, with complex hierarchical and nested natural and human activity environment predictors that might make good sense in a particular context. I've just been impressed with the flexibility and utility of the Bayesian modeling approach, and I urge readers who want to know more and build more realistic models to check out McElreath (2020)

**Chapter Five Key Points**

- Most retrospective analyses of bioassessment datasets to examine temporal and other factors require **considerable investment of time to get the dataset into shape for re-analysis**.
- For the most part, **variation through time in the natural and human activity environment**, and the effect of both on the biota, **have been ignored** in bioassessment programs.
- If change through time is an important question as part of a bioassessment program the **sampling design** has to reflect its importance.
- Adding **year of sampling of the biota** and **year of characterization of the natural and human activity environment** could help discern temporal patterns but is limited by the original design of the bioassessment program.

**Chapter Five R Scripts**

GLBEAST

**GLBEAST 5_1 Diagnosis Model Richness w Time.R—Building a diagnosis model that includes sampling year as a candidate predictor of taxon richness.**

```
library(rethinking)
library(tidyverse)
# From McElreath 2020 pp78-91

# Prepare data for fitting
GLBEAST<-readRDS("./GLBEAST/Datasets/GLBEAST RCA2_trim.Rds")
GLBEAST$log_Fetch<-log(1+GLBEAST$Fetch)
GLBEAST_Ref<-subset(GLBEAST,StatusRCA=="R")
GLBEAST_Test<-subset(GLBEAST,StatusRCA=="T")

# Create predictor Catchment Area varriables that will maintain meaningful y-intercepts
xbar_log_Fetch<-mean(GLBEAST_Ref$log_Fetch)
GLBEAST_Ref$delta_log_Fetch<-GLBEAST_Ref$log_Fetch - xbar_log_Fetch
GLBEAST_Test$delta_log_Fetch<-GLBEAST_Test$log_Fetch - xbar_log_Fetch

xbar_Fetch<-mean(GLBEAST_Ref$Fetch)
xbar_Year<-mean(GLBEAST_Ref$Year)
GLBEAST_Ref$delta_Fetch<-GLBEAST_Ref$Fetch - xbar_Fetch
GLBEAST_Test$delta_Fetch<-GLBEAST_Test$Fetch - xbar_Fetch
GLBEAST_Ref$delta_Year<-GLBEAST_Ref$Year - xbar_Year
GLBEAST_Test$delta_Year<-GLBEAST_Test$Year - xbar_Year


# Fit null model with optional starting values for optimization
flist<-alist(
  Richness ~ dnorm(mu,sigma), # This is the model
  mu ~ dnorm(20,10), # This is priors for mu
  sigma ~ dunif(0,10) # This is priors for sigma
)
# startvals<-list(
#  mu=mean(GLBEAST_Ref$Richness),
#  sigma=sd(GLBEAST_Ref$Richness))
EH0_Null <- quap( flist, data=GLBEAST_Ref)
precis(EH0_Null,depth=2,digits=4)


# Fit a model with common slope for Fetch and separate intercepts for each group -
EH3_GreatLake_log_Fetch<-quap(
  alist(
    Richness ~ dnorm(mu,sigma),
    mu<-a_fit[GreatLake] + b*(delta_log_Fetch),
    a_fit[GreatLake] ~ dnorm(20,10),
    b ~ dnorm(1,1),
    sigma ~ dunif(0,10)
  ),
  data=GLBEAST_Ref)
precis(EH3_GreatLake_log_Fetch,depth=3,digits=4)

# Add Year to the best Reference Condition model
EH3_GreatLake_log_Fetch_Year<-quap(
  alist(
    Richness ~ dnorm(mu,sigma),
    mu<-a_fit[GreatLake] + b_EF*(delta_log_Fetch) +b_Year*(delta_Year),
    a_fit[GreatLake] ~ dnorm(20,10),
    b_EF ~ dnorm(1,1),
    b_Year ~ dnorm(1,1),
    sigma ~ dunif(0,10)
  ),
  data=GLBEAST_Ref)
precis(EH3_GreatLake_log_Fetch_Year,depth=3,digits=4)

# Add Year*GreatLake to the best Reference Condition model
EH3_GreatLake_log_Fetch_YearLake<-quap(
  alist(
    Richness ~ dnorm(mu,sigma),
    mu<-a_fit[GreatLake] + b_EF*(delta_log_Fetch) +b_Year[GreatLake]*(delta_Year),
    a_fit[GreatLake] ~ dnorm(20,10),
    b_EF ~ dnorm(1,1),
    b_Year[GreatLake] ~ dnorm(1,1),
    sigma ~ dunif(0,10)
  ),
  data=GLBEAST_Ref)
precis(EH3_GreatLake_log_Fetch_YearLake,depth=3,digits=4)

# Now compare the models - smallest WAIC wins!
compare(EH3_GreatLake_log_Fetch,EH3_GreatLake_log_Fetch_Year,EH3_GreatLake_log_Fetch_YearLake,
func=WAIC)
plot(compare(EH3_GreatLake_log_Fetch,EH3_GreatLake_log_Fetch_Year,EH3_GreatLake_log_Fetch_Year
Lake))
```

## References

Bailey, R.C., Reynoldson, T.B.: Can datasets from long-term biomonitoring programs detect climate change effects on stream benthos? Sci. Prog. **106**, 368504231219335 (2023)

Bailey, R.G., Ropes, L.: Ecoregions: the ecosystem geography of the oceans and continents. Springer, New York (1998)

McElreath, R.: Statistical Rethinking: A Bayesian Course with Examples in R and STAN, 2nd edn. Chapman & Hall/CRC (2020)

# Chapter 6
# Epilogue: Making and Keeping Freshwater Ecosystems Healthy

I've worked on this book for quite a while, and anybody who has written a monograph knows that it's a non-linear process that meanders in and out of your professional and personal life, sometimes getting your full attention for a few precious days or weeks, but often cast aside for more pressing, day-to-day responsibilities. Motivation is lacking on some days, or even for a few months. But like most people who work on the science of bioassessment, the desire to contribute to the conversation about how to assess the degradation and rehabilitation of freshwater ecosystems has kept bringing me back to this work.

In Chapter 1, I made the argument that the logic of bioassessment is simple…what lives (and doesn't live) in an ecosystem reflects the natural and human activity environment of the ecosystem. Many bioassessment programs have not paid enough attention to characterizing the environments of their target ecosystems, and have struggled to find the appropriate way to distinguish Reference and Test sites. Ironically, for many millennia, Indigenous people have maintained a more intimate relationship with the land and water than the rest of us, often with a keener sense of when and how human activity is changing the world they are a part of in important ways. The approach I am suggesting here does not replace what is sometimes called Traditional Ecological Knowledge, but I think can perhaps complement it to help Indigenous and non-Indigenous stewards of the ecosystems make better choices.

My main message has been to pursue clarity in purpose above automatic execution of the usual way bioassessment programs are designed, executed, analysed, and interpreted. I am desperately *not* in pursuit of a new analysis method that will replace all of the others developed in the last 50 years … my goal is simpler than that … I just want bioassessment researchers and practitioners to think a little more about what exactly they're doing and the choices they are making. For me, the conceptual model is key… I expect criticism of my simple models and the descriptors of the natural and human activity environment I've chosen. But I purposely stayed away from long-winded examples of dozens of candidate predictors and several ways of characterizing the biota along with multiple techniques of doing the modeling

 
R. C. Bailey, *Rethinking Bioassessment of Freshwater Ecosystem Health*,
https://doi.org/10.1007/978-3-032-12427-2_6

and bioassessment. I wanted to keep what's essential clear at the risk of having the approach rejected because colleagues think that I'm just not capturing enough of the nuance to improve on what's been done in the past.

I close with a combination of excitement about sharing what I think is a powerful approach to bioassessment, and humility in knowing its, and my, limitations. I look forward to colleagues old and new, and some I have not yet met, sharing their ideas and constructive criticisms. Those of us who work in this area share a passion, and a responsibility, for freshwater ecosystems, that endures.

MIX
Papier aus verantwortungsvollen Quellen
Paper from responsible sources
FSC® C105338

If you have any concerns about our products,
you can contact us on
**ProductSafety@springernature.com**

In case Publisher is established outside the EU,
the EU authorized representative is:
**Springer Nature Customer Service Center GmbH**
**Europaplatz 3, 69115 Heidelberg, Germany**

Printed by Libri Plureos GmbH
in Hamburg, Germany